Kant's Organicism

Kant's Organicism

Epigenesis and the Development of Critical Philosophy

JENNIFER MENSCH

The University of Chicago Press   Chicago and London

The University of Chicago Press, Chicago 60637
The University of Chicago Press, Ltd., London
© 2013 by The University of Chicago
All rights reserved. Published 2013.
Paperback edition 2015
Printed in the United States of America

24 23 22 21 20 19 18 17 16 15   2 3 4 5 6

ISBN-13: 978-0-226-02198-0 (cloth)
ISBN-13: 978-0-226-27151-4 (paper)
ISBN-13: 978-0-226-02203-1 (e-book)
10.7208/chicago/9780226022031.001.0001

Library of Congress Cataloging-in-Publication Data
Mensch, Jennifer.
  Kant's organicism : epigenesis and the development of critical philosophy / Jennifer Mensch.
    pages. cm.
  Includes bibliographical references and index.
  ISBN 978-0-226-02198-0 (cloth : alk. paper) — ISBN 978-0-226-02203-1 (e-book)   1. Kant, Immanuel, 1724–1804.   2. Philosophy of nature.   I. Title.
  B2799.N37M46 2013
  193—dc23   2012043133

♾ This paper meets the requirements of ANSI/NISO Z39.48-1992 (Permanence of Paper).

For Zoe, whose flourishing has lived up entirely to the promise of her name.

Systems seem to be formed in the manner of lowly organisms, through a *generatio aequivoca* from the mere confluence of assembled concepts, at first imperfect, and only gradually attaining to completeness, although they have one and all had their schema, as the original germ, in the sheer self-development of reason. Hence, not only is each system articulated in accordance with an idea, but they are one and all organically united in a system of human knowledge, as members of one whole, and so as admitting of an architectonic of all human knowledge.
CRITIQUE OF PURE REASON, A835/B863

Contents

Preface ix

Introduction: Kant's Organicism   1

1   Generation and the Task of Classification   16

Mechanism and the Principle of Life 16
Leibniz's Organic Machines 28

2   Buffon's Natural History and the Founding of Organicism   35

Hales and the Physiology of Plants 35
Buffon the French Newtonian 38
Maupertuis, Buffon, and the Problem of Form 43
Natural History and the History of Nature 47

3   Kant and the Problem of Origin   51

Kant's Eclecticism 51
Matter and Cosmos 53
The Spectacle of Life 60

4   The Rebirth of Metaphysics   70

A Philosophy Is Born 70
From Original Acquisition to the Epigenesis of Knowledge 80
Concepts and Objects: Kant's Letter to Herz, 1772 83

5   From the Unity of Reason to the Unity of Race   92

   The Unity of Reason 92
   The Unity of Race 95
   A Germ of Reason and a Germ for Race 104

6   Empirical Psychology in Tetens and Kant   110

   Epigenesis and Evolution in Tetens's Philosophical Essays 110
   From Empirical Psychology to a Transcendental Theory of Imagination 114
   Transcendental Philosophy and the Physiology of Pure Reason 119

7   Kant's Architectonic: System and Organism
    in the *Critique of Pure Reason*   125

   The Doctrine of Method: The Bauplan *of the System* 125
   The Transcendental Deduction: The Bauplan *at Work* 131
   Organic Logic: A Cautionary Tale 140

   Epilogue: A Daring Adventure of Reason   146

   Notes 155   Bibliography 219   Index 239

Preface

Immanuel Kant has maintained an enduring intellectual presence through his works on morality, reason, history, and art. He created the first university courses on physical geography and anthropology, and throughout his career he taught logic and metaphysics alongside courses discussing everything from taste to table etiquette. It is estimated that by the time Kant died there were already well over three thousand published pieces devoted to his work, and even as Kant's general influence waned toward the end of the nineteenth century, new currents emerged such that "Neo-Kantianism" came to describe a number of schools in philosophy. Kant's moral theory remains to this day a pillar of classical ethics and a centerpiece in contemporary bioethical discussions of autonomy and patients' rights, and he continues to hold interdisciplinary appeal across various fields of law, science, and the humanities. In recent times, Kant has attracted added attention from historians of science and critical race theorists for his work in natural history and, as some have it, for his invention of the concept of race. It is such long-standing and widespread interest in Kant's work, interest stemming from all manner of intellectual backgrounds and any number of investigatory goals, that has made Kant one of the most widely discussed authors in the history of ideas.

Given the very breadth of Kant scholarship, it is perhaps useful to locate this book, at least in a topographical vein, within its appropriate region. *Kant's Organicism* starts by tracing the history of the life sciences as Kant would have come to know them, focusing especially on those philosophers and life scientists whose works directly engaged Kant during his intellectually formative years. Once Kant's connection to the life sciences has been established, the remainder of this book moves to an examination of the exact nature of the influence of these sciences on the emerging critical system. When viewed from the perspective of the life sciences in this manner, Kant's theoretical philosophy becomes reframed as a philosophical project whose development was deeply influenced by the rise of organicism, a movement that arose in the wake of developments in natural history and helped shape fields as diverse as science, literature, politics, and philosophy. The general argument for Kant's organicism is outlined in the introduction, with the details left to be developed in the chapters that follow.

There are a great many people to thank when one writes a book, and I am glad for the opportunity here to express my gratitude for all of the help and support I received along the way. Tracking down obscure historical references is a time-consuming endeavor, and I was fortunate throughout to have had the tireless help of Claudia Villafranca from Pennsylvania State University's Interlibrary Loan division. Special thanks go to Mary Terrall for not only generously sharing her private notes on Maupertuis's Baumann thesis but also pointing me toward Berlin as a resource for this manuscript in the first place. Peggy Price, curator of Special Collections at the University of Southern Mississippi, patiently went through volumes of the German edition of Buffon's *Histoire naturelle* in search of references for me. Eric Watkins gave special help with translation questions related to Kant's scientific works dating from his earliest precritical writings; Holly Wilson was intrepid in resolving a number of problems, dating and otherwise, regarding Kant's anthropological essays and lectures; and Robert J. Richards provided both feedback and guidance concerning the relationship between Blumenbach and Kant. Three of my colleagues in the Department of Philosophy are to be especially thanked for their continuous support and encouragement regarding the project, Robert Bernasconi, Brady Bowman, and Mark Fisher. My thanks also to Peter Giannopoulos, who lent his talent and energy to the book in its final stages by preparing the bibliography.

When I began this book, I had already been lecturing on Kant for a good number of years, and it is a pleasure to express my appreciation here for the Kant scholars whose teaching and work first inspired me as their student and whose influence has continued to affect me as a professor and scholar. For this I want to thank Rudolf Makkreel, Eckart Förster, Manfred Kuehn, Hoke Robinson, and Mark Timmons. *Kant's*

Organicism also benefited from the readers' comments made by John Zammito and Günter Zöller; I am grateful for the time and energy they put into their reviews, and I hope they will feel that the book has been improved as a result. At the University of Chicago Press David Brent has been ideal as both an editor and overall supporter of the project; his editorial associate, Priya Nelson, has been in equal measure efficient, friendly, and helpful in steering the book through all its various stages from review to production. Finally, I am especially grateful for George Roupe's careful work and thoughtful suggestions when copyediting the final manuscript for Chicago.

I received a great deal of support while writing this book from my family, including, of course, my dog Ollie, who stayed by my side during every minute that I worked on it. My mother and father, Josephine and James Mensch, and my brother and sister, Joshua and Jessica Mensch, have been as good as it gets for unconditional support, encouragement, and general partisanship on my behalf during the entire process from beginning to end. My daughter, Zoe Mensch Schmidt, has been both patient beyond her years and full of good suggestions for wrapping up the project a bit more speedily than it has been, reminding me with some significance on more than one occasion that "staples have always worked well" for her when putting the finishing touches on one of her own books. My greatest thanks of all go to my husband, Dennis J. Schmidt, who not only read through and edited the manuscript three times from beginning to end but kept the house and everything in it, not least including me, sane, organized, and happy; for that and more, Denny, thank you.

A portion of chapter 1 appeared previously in slightly different form as "Understanding Affinity: Locke on Generation and the Task of Classification," *Locke Studies* 11 (2011): 49–71. The image used in the conclusion is a reproduction of the title page of Francis Bacon's *Instauratio Magna*, held by the Rare Books Collection at the University of Chicago Library; my thanks to the staff at the Special Collections Research Center for their help in procuring the image and granting permission for its use.

INTRODUCTION

Kant's Organicism

This book is oriented by the conviction that Kant should be fitted into a framework that has begun to take shape in a number of fields when it comes to thinking about the mid- to late eighteenth century, a framework that can be called something like "organic thinking" or, better yet, "organicism." Organicism can be defined by its view of nature as something that cannot be reduced to a set of mechanical operations. The stage for organicism was historically set by investigations into the connected concerns of natural history and embryogenesis, investigations leading to inevitable conclusions regarding nature's vitality and power. And while historians of science have long understood the centrality of these investigations to the late eighteenth century as a whole, it is increasingly the case that disciplines outside of science are now producing studies of the period along similar lines. At this point there are numerous accounts of "epigenesist poetry" and "epigenesist literature"; there are political theorists who speak of "Enlightenment vitalism," and the utopian literature of the period is said to employ "the language of epigenesis" when describing the ideal society. Indeed, in light of all this activity one cannot help but reach the conclusion that the latter half of the long eighteenth century is a period best defined by its organicism. For organicism, used interchangeably with "epigenesis," a term borrowed from embryological theory, seems best to describe the response by science and art, in politics and literature, when grasping the problems and possibilities of an irreducibly living nature.[1]

INTRODUCTION

Now it has become customary for literary critics and historians alike to pay passing tribute to Kant's role in this narrative, a tribute paid almost without exception to Kant's third *Critique*, the *Critique of Judgment*, a book devoted to an investigation of nature and art. Kant's language of "reflective judgment" and his appeal to transcendental principles as heuristic guides for "orientation" were modes of epistemic caution that were for the most part ignored as the possibilities for connecting teleology and mechanism and for discovering freedom within nature and art were taken up instead by Kant's successors. There are in fact numerous points of contact between the *Critique of Judgment* and the Romantic science that would follow, but I want to investigate the degree to which Kant—and not just Kant as he was appropriated through the third *Critique*—can be located within a period defined by its organicism in order to discover in what manner Kant too would be attracted to the model offered up by "epigenesis" for thinking about questions of origin and generative processes in general. For it is my sense that epigenesist models had a significant role to play for Kant's theory of cognition, for what one might even go so far as to describe as his epigenesist philosophy of mind. And I believe that it is in fact only through attention to this influence, to seeing Kant's organicism as it were, that we can both make sense of the transcendental deduction at the heart of Kant's theory of cognition and discover the means by which his work in natural history can be meaningfully integrated into the critical system as a central part of the whole.

Before turning to Kant, however, it is worth pausing briefly to rehearse the general state of the life sciences as Kant would have first come to appreciate them in the 1750s and 1760s. By 1772 Thomas Ramsay could write that "natural history is, at present, the favourite science over all Europe, and the progress which has been made in it will distinguish and characterize the eighteenth century in the annals of literature."[2] Answering the question as to why natural history would achieve the kind of popularity it would enjoy well into the nineteenth century would take us too far afield, but at least a few of the contributing points can be made so far as these set the stage for organicism. By midcentury, for example, serious challenges had been laid down against the reigning theory of generation and indeed the general portrait of organic life as a whole. For much of the century before this, those working in the life sciences could be roughly divided into experimenters and systematists. This division is important to notice, since it is precisely the convergence of what had been parallel tracks, of experimentation with organic processes on the one hand and of the systematic classification of individual organisms on the other, that both established natural history as something that Ramsay would have recognized and became a basis for challenging the received view.

Until the 1740s, theories of generation, and of embryogenesis in particular, were oriented by a belief in the preexistence of all biological organisms. The position sounds fantastic today, but at the time, there were good reasons for its central role in biological theory. The notion that God had created every individual at the beginning of history relieved naturalists of the need to explain the means by which organisms might manage the imposition of form and force on an otherwise lifeless matter; that material being was indeed lifeless apart from God's agency had firm support from post-Reformationist schools of thought. Preexistence made room, moreover, for the increasingly secluded mechanical philosophy when it came to the explanation of organic generation. No one had been convinced by the Cartesian analysis of generation as a form of fermentation, and thus there was almost a sense of relief when mechanism assumed once more an important role to play for explaining the processes of nutrition and growth in the expansion of the previously formed yet submicroscopic individual. It was in fact the microscope that, more than anything else, lent credibility to the theory once experimenters discovered what they took to be miniature homunculi encapsulated in the "spermatic worms" seen by Leeuwenhoek in the late 1670s. Finally, it was a matter of particular convenience for the systematists to endorse preexistence so far as it ensured that for all the difficulties facing taxonomy the objects of that science would remain stable. As Linnaeus suggested, it might be tricky to determine whether the mulberry belonged with the nettles, but at least one could be sure that mulberries as a species were fixed.

The tide began to turn against preexistence theories in the 1740s, starting with Abraham Trembley's spectacular discovery of the freshwater hydra. This polyp appeared to be infinitely plastic with respect to its possibilities for regeneration. It could be sliced, severed, turned entirely inside out: in every case the hydra either regenerated the lost part, generated a second individual, or, in the last instance, simply grew a new outside altogether. The impact of this discovery cannot be overestimated for its revolutionizing effect on the life sciences. Questions poured out as a consequence of this discovery: How could preexistence theory explain this capacity? How, in this instance, could one insist on the lifelessness of the animal-machine? It hardly helped matters to note the problem of categorizing the polyp altogether, so far as it seemed to be essentially a plant with a stomach. Problems in classification had in fact begun to multiply as botanists in particular complained of the difficulty in fitting their observations to Linnaeus's system, and categories assigned to indeterminate species thus slowly began to overshadow the so-called pure lines. In the late 1740s, Pierre Louis Maupertuis, the newly elected president of the Berlin Academy of Sciences, began to collect records that he would publish on a family known for its many cases of polydactyly. If, as those records indicated, a trait could be passed on by both female and male members of the family, the basic tenets of preexistence theory had to be wrong: generation must be an active process, one clearly requiring the contribution of both mother and father in the production of an embryo. Against this kind of evidence, it almost seemed beside the point to wonder what God would have had in mind when preforming deformities such as those experienced by the family of polydactyls.

Hybrids, hydras, "monsters": these were all certainly on Georges Buffon's mind as he sat down to begin composing what would eventually grow to be some three dozen volumes on natural history. The first three volumes, appearing together in 1749, were almost immediately translated into German, and Buffon's significance in laying the groundwork for the organic view and the German strain of organicism in particular is clear. Buffon had correctly assessed the central problem facing the taxonomical system as one based on a fundamentally inaccurate view of both nature and knowledge. Nature was not rigidly demarcated along the lines proposed by the taxonomists, nor should one ever hope to completely grasp its manifold principles and operating causes when assessing its effects; at best, according to Buffon, one could adopt the strategy of a kind of game theory, using probabilities as a guide when determining the contours of our species maps. Buffon understood the consequences of his position. If research into organic processes revealed natural agency, then natural history would have to redefine itself as a discipline devoted to the *histories* of living things; it would need to commit itself, in other words, to the principle that nature was susceptible to change. And the first site of this capacity for change was embryogenesis. Devoting almost the entirety of volume 2 to the problem of generation, Buffon made development the basic biological process, the key to understanding natural history as a science of living nature. For it was here, during the composition of the embryo, that change could be affected by environmental factors such as food and climate. Change produced variation, or "degeneration" in Buffon's terms, and it both explained the experience of affinities when viewing varieties and grounded a historical sequence capable of linking, to use one of Buffon's favorite examples, the "proud mouflon" on the mountaintop and the pathetic sheep in the field. It is Buffon, then, who best marks the moment of convergence necessary for the establishment of natural history: the previously parallel investigations into system and process converged in Buffon's natural history to produce both a new view of organic life and the basis for redefining taxonomy as a form of genealogy.

When it came to describing embryogenesis, Buffon relied on something he called an "internal mold"; it marked Buffon's attempt to provide a pseudomechanical explanation of the means by which form could be conveyed to the organic material of an embryo. Sometimes described as "mechanical epigenesis" to distinguish it from its more vitalistic conception, the term "epigenesis" was rapidly appropriated beyond any one theory to represent all positions counter to preexistence.[3] Epigenesis was, however, an old idea. Aristotle had considered the process by which the male imparted the soul—as source of both information and animation—to material provided by the female in terms that would suggest epigenesis to his later readers.[4] Thus in 1651 Harvey understood himself to be following Aristotle when using epigenesis to describe the progressive development of a chicken embryo from homogeneous mass to heterogeneously structured organism.[5] Harvey refrained from speculation regarding the basis of this organizational drive, as did Caspar Wolff, who published experimental results that he took, in 1759, to be evidence of a nutritive life force, a force that he called *vis essentialis*.[6] Wolff's observations suggested a dialectical logic underlying generation, an incessant motion that, in the case of plants, explained development as a back-and-forth motion between fluidity and solids. Epigenesis thus met a need to grasp the power and vitality of nature, but without recourse to the soul or devices such as Buffon's interior molds, it faced an impossible task with respect to the problem of form. As one critic complained, the epigenesist "needs a force which has foresight, which can make a choice, which has a goal, which, against all the laws of blind combination, always and unfailingly brings about the same end."[7] Despite this concern, epigenesis would soon become the common denominator of organicism: a model for literature and politics as much as for Romantic science itself.

Turning to Kant now, one discovers that within two years of Kant's passing the requirements that would allow him to teach, he received special permission to offer a new course, a course that Kant called "Physical Geography," which in outline carefully followed the path taken by Buffon in the first volume of his natural history. It was 1757, and Kant had already established his interest in the problem of origin. His most important works had so far been devoted to questions regarding cosmological origin, with numerous small pieces devoted to geological formation and natural processes associated with the workings of wind, fire, and earthquakes. So it comes as no surprise to learn that Kant kept abreast of debates regarding organic generation as well. On the whole, he took the prospects for any genuine advance in the life sciences to be gloomy. Physics was easily reducible to a set of mechanical causes, but, Kant asked, "Can we claim such advantages about the most insignificant plant or insect? Are we in a position to say: *Give me matter and I will show you how a caterpillar can be created*? Do we not get stuck at the first step due to ignorance about the true inner nature of the object and the complexity of the diversity contained in it?" (1:230).[8] The problem of generation was simply closed off from examination, at least so far as Kant was concerned.

It stands, therefore, as a tribute to the rising prominence of debates over preformation and the epigenesist alternative that the by then well-regarded Magister Kant took the opportunity to review the options as he saw them in 1763. The problem with preformation was that it relied on an essentially supernatural explanation, and recourses to God at this juncture in the history of science were simply no longer compelling. That said, Kant thought that "it would be absurd to regard the initial generation of a plant or an animal as a mechanical effect incidentally arising from the universal laws of nature" (2:114). What was needed was something different, a means of avoiding the supernatural solution even if all of the mechanical accounts of generation had so far failed. Mindful of the need to provide form, Kant emended the epigenesist alternative. Is it possible, Kant asked, that "some individual members of the plant and animal kingdoms, whose origin is indeed directly divine, nonetheless possess the capacity, which we cannot understand, to actually generate [*erzeugen*] their own kind in accordance with a regular law of nature, and not merely to unfold [*auszuwickeln*] them?" (2:114).[9] Kant's suggestion, in other words, proposed a compromise. Form was indeed supernaturally conceived, but while this generically maintained the stability of the species lines, the work of generating individuals actively belonged to nature. And the distance epigenesis had come from Buffon's account was clear not only from Kant's direct dismissal of that position as an "entirely arbitrary invention" but from the emphasis placed on a specifically nonmechanical account of organization.[10]

At this point in history there were a number of ways in which the term "epigenesis" was used. Above all, epigenesis referred to the production, the actual generation, of something new. And it was in this sense that detractors could link the notion to older, discredited claims regarding the spontaneous generation of flies and so on. Epigenesis, so far as it was identified with a theory like Buffon's, emphasized the fact of joint inheritance and so was associated with an account of "blending." Also in play were the two earlier accounts: Harvey's observationally based definition of epigenesis as the development of increasingly heterogeneous structures from out of an initially homogeneous mass and Aristotle's discussion of the imparted soul.

Kant was familiar with all of these uses. In his lecture course on metaphysics he contrasted the relative advantages offered by a preformation theory compared to epigenesis for couples, so far as epigenesis would require careful consideration of what the blended progeny might be like (17:416). Kant also regularly found opportunity to criticize Aristotle's account as fundamentally absurd given the impossibility of dividing or sharing a simple substance like the soul (17:672, 18:190, 18:429, 28:684, 23:106–107). And although he considered the possibility that biological epigenesis might offer a real alternative to mechanical models of generation (17:591), Kant worried over the difficulty of finding a principle that would be capable of explaining the stability of epigenetic development against potentially altering sources presented by the environment (18:574). Kant's final position regarding organic embryogenesis would sound close to the position that he had first outlined in 1763. Thus in 1790 Kant would describe epigenesis as akin to a system of "generic preformation" according to which "the form of the species [is] preformed *virtualiter* in the intrinsic purposive predispositions [*Anlagen*] imparted to the stock" (5:423), a position to be preferred so far as "it minimizes appeal to the supernatural, and after the first beginning leaves everything to nature" (5:424).[11] Two senses of epigenesis remain: the sense of it as a type of spontaneous generation and Harvey's technical description of development as a movement from undifferentiated unity to an interconnected whole of diversely functioning parts. It was these two models of biological epigenesis that would prove to be most influential for Kant's *metaphysical* account of cognition, an influence that would in turn clarify Kant's subsequent investigations into natural history.[12]

Starting in the mid-1760s Kant's attention began to turn away from concerns regarding cosmological and biological origin and toward a constellation of problems surrounding the basis of knowledge and, in particular, the origin of ideas. The problems were pressing. In metaphysics and natural science alike confusion reigned, according to Kant, as the result of insufficient attention to the bases upon which claims were being made and the careless, free-flowing use of vocabularies across the sciences. It was simply wrong to take concepts borrowed from physics, concepts like attractive and repulsive forces for example, and apply them uncritically when attempting to explain something like the metaphysical connection between body and soul. And the attempt in the life sciences to establish something like Wolff's *vis essentialis* as an actual "principle of life" or soul within matter was no different (28:275, 283). In each case a force was asserted to explain an effect that might very well be acknowledged to exist but that resisted all mechanical attempts at explanation nonetheless (2:331). Mechanical explanation, as Kant came increasingly to believe, was the only kind available with respect to determinate knowledge of nature. Thus while Kant ultimately took generic preformation to offer the most defensible response to the problem of generation, this was an endorsement with a caveat. So long as the keys to organic processes resisted mechanical reduction, they simply could not be known with the kind of certainty afforded the nonbiological sciences of mechanics and physics. Biology could not, therefore, be realized as a complete science, and all hypotheses regarding organic formation and natural history at large would have to remain heuristic at best.

This was not the case, however, for investigations into the cognitive processes underlying the generation of knowledge. Once Kant declared metaphysics to be henceforth known as a science of the extent and limits of knowledge, the first task was to examine the basis of its claims. Taking stock of his options, Kant considered the alternatives offered by Leibniz and Locke. Leibniz, no less than the preformationists, on Kant's view, relied on a supernatural explanation when it came to the origin of ideas. Locke's insistence on a sensible basis, however, failed to appreciate the role played by mental reflection when generating concepts that were irreducible to sense data (28:233). In contrast to either of these positions, Kant was ready by 1771 to describe his own position as "epigenetic." The "real principle of reason," Kant now argued, rests "on the basis of *epigenesis* from the use of the natural laws of reason" (17:492). Only one year before, Kant had had to content himself with tracing intellectual concepts back to what he had then described as their "original acquisition" via attention to the lawful workings of the mind. While this had allowed Kant to avoid the alternatives of concepts that were either sensible or innate, the explanation of just what was meant by "original acquisition" was missing. By subsequently identifying epigenesis as the model for cognition, Kant seems, to borrow Darwin's phrase, to have at last found "a theory by which to work."[13]

When Kant began work in earnest on the series of investigations that would lead to the publication of the *Critique of Pure Reason* in 1781, he stopped publishing entirely in the subject matter of the *Critique*. It is thus a matter of special significance to see that Kant's main publications during this period were in natural history, for only these could be conceptually linked to the somewhat parallel investigations into the bases of cognition. Kant's single appearance in print between 1770 and 1775 was the review of an Italian anatomist's discussion of the structural similarities between humans and animals, similarities that, in the anatomist's view, led to the conclusion that all manner of ailments resulted from humanity's "unnatural" state of two-footedness (2:421–425). In his response, Kant deferred to the medical expertise of the anatomist, but suggested, nonetheless, that a fundamental difference remained so far as humans alone contained "a germ of reason" (*ein Keim von Vernunft*), which if developed (*entwickelt*) would destine them for society; it was a point that Kant would continue to raise against Moscati, named or not, in subsequent lectures on physical geography and anthropology. During the remainder of the decade Kant would gradually come to realize the full consequences of what it might mean to have an epigenesist conception of mind, a mind that, like the organism itself, would have to be viewed as operating according to a kind of reflexive or organic logic according to which its unity must be viewed as both cause and effect of itself.

Until the middle of the 1770s Kant took the generation of representations to be something requiring a juggling of factors directly parallel to those in play when considering organic generation. There had to be something regular, like a set of rules, guaranteeing uniformity of production. There had to be material content, and there had to be some kind of force, something capable of putting the parts together according to the rules. Finally, there had to be something capable of maintaining the unity, if not the identity, of the whole—a simple enough set of requirements perhaps, but the work, as usual, lay in the details. The immediate challenge concerned the specific connections between the various mental faculties in play—the faculty of understanding as home to the rules, sensibility as provider of material content, and *eine bildende Kraft*,[14] a formative power capable of connecting the material to the rules—a challenge exacerbated by Kant's commitment to a solution relying on neither supernaturally preformed ideas nor the empiricists' appeal to sense. The intellectual intuition of innate ideas simply smacked of "lazy philosophy," according to Kant, while the empiricists invited a skepticism that could only damage sciences grown increasingly reliant on induction.

By 1775 Kant had made good progress. Intellectual concepts—concepts like "substance" and "causality"—were now said to be based on rules for the logical positioning of sense data. Logical positioning explained how judgments were formed; indeed it defined cognition as a whole so far as cognition was now said to "consist in judgments" (17:620). Experience would be lawful and skepticism thereby avoided to the extent that cognition predetermined it according to the rules of logical positioning. Kant had in fact already been clear since 1770 on the fact that truth could be won so far as attention was paid to the rules for constructing appearances, rules that amounted to determining the logical connection between predicates in a judgment. The advance since then was to identify *concepts* with the rules for logical connection (17:614). It was from these rules that Kant could understand the epigenesis of concepts from the use of the natural laws of reason. But what was the status of these laws and rules? Were they in fact as preformed as the *supernaturally* preformed germs generically maintaining the species lines? Kant's notes during this period concentrate on the process of judgment formation itself, with page upon page devoted to working out the steps between a "principle of disposition" (*Disposition*) or "aptitude" (*aptitudo*) for organization (17:656) and the "exposition" (*exposition*) of this organization as a kind of exhibition, expounding, or realization of the rules themselves (17:643, 644, 648, 656, 660, 662).[15] This exposition of the rule, a representation of logical connection, generated unity, according to Kant, since the connecting of predicates in a judgment was precisely what unified an aggregate of sensation into a meaningful system of representation.

It was at precisely this stage in Kant's reflections that he took up the option of attaching a short essay to his regular set of course announcements for the 1775–1776 school year. It would be the last time Kant would publish this kind of advertisement, this time to announce that the course on physical geography would be taking up a question of increasing interest in natural history, namely, the explanation of race. Polygenesists had been maintaining that races represented distinct lines of creation, that they were in fact so many different kinds or species. Kant, following Buffon's adoption of interfertility as the only suitable criterion for determining species, argued instead for monogenesis. The job for naturalists interested in explaining the grounds of racial difference was therefore twofold, explaining the causal basis of such adaptation—for Kant took the generation of racial characteristics to have originally been an adaptive response to environmental conditions—and explaining the patterns of geographic isolation with respect to these adaptations, explaining, in other words, why similar occasioning causes like high heat and aridity did not seem to have produced similar races in all such locations with those characteristics.

Our interest concerns Kant's explanation of adaptation so far as it returns us to the language of germs and dispositions. By this point preexistence theorists had had to respond to discoveries like those regarding the regenerative possibilities of the hydra. And the most successful response, by far, had been put together by the Swiss naturalist Charles Bonnet. Bonnet had argued that organisms contained innumerable germs, germs containing the imprint of the species, and Kant seems to have had a similar strategy in mind when discussing the basis of biological adaptation.[16] According to Kant, the only way to explain environmental adaptation was to suppose the preexistence within species lines of "germs" for new parts and "natural predispositions" for proportional changes to existing parts. Kant took the case of birds as his first example in the course announcement. As he explained it, "In birds of the same kind which yet are supposed to live in different climates there lie germs for the unfolding of a new layer of feathers if they live in a cold climate, which, however, are held back if they should reside in a temperate one" (2:434). But how was one to understand the existence of such spectacular provisions for adaptation? Surely neither chance nor mechanical laws could explain the existence of germs purposed for the possibility of an organism's adaptive needs. "The human being," Kant continued, "was destined for all climates and for every soil; consequently, various germs and natural predispositions had to lie ready in him to be on occasion either unfolded or restrained, so that he would become suited to his place in the world and over the course of the generations would appear to be, as it were, native to and made for that place" (2:435). What Kant wanted was a lawful basis for adaptation. The existence of germs purposed for human survival across climate and geography seemed to explain both the fact of adaptation and its inheritance. Like Harvey's definition of epigenesis as the movement from homogeneous unity to increasingly distinct parts, the natural history of the human species could be viewed similarly with monogenetic unity securing phyletic connection and germs providing the rules for subsequent differentiation. But this kind of conclusion, as always with biological explanations, carried a caveat. So long as the actual histories of species remained unknown, natural history as a genealogical enterprise would fail to offer precisely that set of laws required for its establishment as a science. The "physical system for the understanding" (2:434), as Kant called it in 1775, would never be realized as an empirical science.

Returning to his work on the *Critique of Pure Reason*, Kant was ready to make a distinction, one that would prove to have a deep conceptual impact on the critical project as a whole. There had to be different grounds for unity in cognition: the rule-based unity of judgments at the heart of representation, and the unity of reason itself—in Kant's words, a "unity of experience" on the one hand and the "unity of *the self-determination of reason* with regard to the manifold of the unity of rules or principles" on the other (17:707–709, italics mine). By describing the unity of reason as a case of "self-determination" Kant had finally located an epigenetic beginning, an origin that was neither supernatural nor empirical but spontaneous. And it was only in the vein of something that could be metaphysically conceived as *self-born* that the unity of apperception could be subsequently referred to as "pure spontaneity" or as "transcendentally free." The rules and intellectual concepts responsible for generating a unified experience would subsequently be described as having been themselves generated, as a set of diversely functioning parts, from out of reason itself. Rather than lying like preformed germs and dispositions, the rules would operate, therefore, like *emergent* properties,[17] constructing experience at the same time that they gave definition to spontaneity itself, realizing or "perfecting" it through their lawful operation. Thus while the unity of reason could be conceptually distinguished from the unity of rules for constructing experience, like an organism, cognition functioned as a set of parts whose thoroughgoing connection realized unity even as the grounds of that unity preceded it. This was a different logic at work than that driving the discursive logic of judgment formation; it was a reflexive logic according to which the unity of apperception was both cause and effect of itself, or, as Kant would put it in another context, both author of and subject to its own laws.

The *Critique of Pure Reason* finally appeared in 1781. It was a book whose energies were divided between attention to the positive account of rules for coherent experience and the negative work of outlining reason's capacity for illusion in its desire to push past the boundaries it had itself set as the ground of experience. The necessity ascribed to the rules for experience became a matter of *genealogy*, as Kant now described the connection between unity of rule and unity of apperception on the basis of their organic affinity. "How," Kant asked, "are we to make comprehensible to ourselves the thoroughgoing affinity of appearances, whereby they stand and *must* stand under unchanging laws?" (A113). Kant's answer lay in neither the kind of "special affinity" affirmed by Leibniz and responsible for connecting innate ideas and intellectual intuition nor the "natural affinity" thought by Hume to form the basis of laws for imaginative association.[18] Organic affinity, in contrast to either of these accounts, secured necessity or lawfulness in experience so far as the rules for connection had their "birthplace" in apperception (A66/B90). "The objective ground of all association of appearances," Kant now declared, "I entitle their *affinity*. It is nowhere to be found save in the principle of the unity of apperception, in respect of all knowledge which is to belong to me" (A122). This was Kant's response to skepticism: rules guaranteed the coherence of experience, and the unity of apperception secured the origin and thereby the legitimacy of the rules. "Our skeptical philosopher," Kant explained, ignored the genealogy of our concepts or rules and thus "proceeded to treat the self-increment of concepts [*diese Vermehrung der Begriffe aus sich selbst*], and, as we may say, this self-birth [*die Selbstgebärung*] on the part of our understanding (the same as of our reason), without impregnation by experience [*ohne durch Erfahrung geschwängert zu sein*], to be impossible" (A765/B793). Only the "self-birth" of reason or, as Kant would later add, the "epigenesis of reason" (B167) could finally secure the coherence of experience. Kant's transcendental deduction, where "deduction" represents a term borrowed from the legal work to determine rightful *inheritance*, could not, therefore, have been more aptly named given the vocabularies of origin and birthright at play.

With the *Critique of Pure Reason* in place, Kant was able to return with greater clarity to natural history. Reviewing Johann Herder's attempt to avoid both preexistence and mechanism in his appeal to a "genetic force" at the basis of adaptations, Kant was ready to agree,

only with this reservation, that if the cause organizing itself *from within* were limited by its nature only perhaps to a certain number and degree of differences in the formation of a creature . . . then one could call this natural vocation of the forming nature also "germs" or "original predispositions" without thereby regarding the former as primordially implanted machines and buds that unfold themselves only when occasioned as in the system of evolution, but merely as limitations, not further explicable, of a self-forming faculty, which latter we can just as little explain or make comprehensible. (8:62–63)[19]

INTRODUCTION

In 1775, Kant's effort to discover the means for lawful adaptation had stopped with the supposition of germs purposed toward the adaptive needs of an organism. By 1784 Kant was prepared to heuristically mirror the language of cognition such that heritable traits, no less than the rules for experience, emerged to limit and therefore realize, constrain and thereby form, a freely exercised power of life.

Without a mechanical explanation of necessary inheritance, Kant turned to a basic tenet of the first *Critique*, namely, that all necessity without exception must have a transcendental ground (A106). The unity of apperception was the transcendental ground guaranteeing the necessary coherence of experience. But in the natural history of the human species no such ground could ever be discovered. It could only be asserted therefore as a transcendental principle: a principle serving as a condition—not for the construction of experience but for the possibility of orientation within it. How can we understand natural history as a genealogical exercise, one capable of providing, in Kant's words, a genuine "archaeology of nature" (5:419)? In the case of the human species it is by asserting the monogenesis of our kind, a phyletic unity requiring the possibility of differentiation from the start given the vagaries of climate and geography (8:99).[20] But the *lawfulness* of original adaptation, the *necessity* of subsequent inheritance, these can therefore only rest on a transcendental principle regarding the unity of our species, a principle we supply as a unifying law of reason, a law that reason gives to *itself* in its investigation of nature as seen through the lens of teleology. This was Kant's solution to the complaint he had first voiced in 1763 regarding the need for recourse to some kind of explanatory principle besides mechanism or God. It was a solution that could yield a productive means for the investigation of nature while still remaining faithful to the limits of our claims.

By setting limits on the use of transcendental principles regarding nature's unity and purposiveness, Kant expressed a note of epistemic caution that would go unheard by his successors. Convinced of nature's vitality, naturalists and philosophers would make use of Kant's work as they saw fit. The most significant transformation of Kant's work concerned the use of transcendental principles themselves, since these tools for *thinking* about nature would be subsequently ascribed to nature itself. This so-called constitutive use of what was meant to be only a transcendental principle for reflective judgment betrayed its lineage as more than an epistemic device, as something that was indeed itself forged out of Kant's synthesis of biological and epistemic concerns. Thus when Goethe described "intuitive perception" as the ability to "see the ideas" at work in nature, he was identifying the archetype as something that functioned both epistemically *and* as the biologically active ground of metamorphosis.[21] This would be the case for Darwin's appeal to "common descent" as well. Descent with modification, the guiding idea behind the theory of natural selection, represented a claim meant not only to orient our investigation of nature but to ground the interconnection of nature itself. Common descent functioned like a transcendental principle so far as it oriented classification toward the search for nature's unity via phyletic lineage between organisms. But it was also more than a mere heuristic by which one could think nature's interconnection; it was the organically real ground of biological affinity, the only basis upon which Darwin could declare comparative anatomy to be "the soul of natural history."[22]

In the end, while Kant's real role in natural history might have operated through the manner in which he was appropriated, his place in the organicism of his time is best secured by his account of the epigenesis of reason, an epigenesis that was far more radical than the one Kant was willing to accord natural organisms via "transcendental principles," and one that locates Kant as a genuine forerunner of investigations into "epigenetics" and the "emergent properties" of genes that are central to discussions of embryogenesis today.

ONE

Generation and the Task of Classification

Mechanism and the Principle of Life

Locke's theory of classification is a subject that has long received scholarly attention. Relatively little notice has been taken, however, of the special problems that were posed for taxonomy by its inability to account for organic processes in general. Classification, designed originally as an exercise in logic, becomes immediately complicated once it turns to organic life, and the aims of taxonomy become thereby caught up with the special problems of generation, variation, and inheritance. Locke's own experience with organic processes—experience gained through his early work in botany and medicine—suggested to him both the dynamism of nature and the necessary artificiality of an a priori system of classification. Locke's attitudes toward nature were not uncomplicated, at times presenting a blend of seemingly opposed commitments. But these were precisely the grounds upon which he could recognize the need to disentangle the epistemic, cognitive aspect of taxonomy from the attempt being made by taxonomists to create a natural system. In the end, it was this disentanglement that would both pave the way for Linnaeus's successful creation of an artificial system of classification and open the door to its subsequent attack by Buffon and his followers. By the middle of the eighteenth century, natural history would be wrested from the hands of taxonomy, but this path could not have been laid without Locke's work to demonstrate the arbitrary nature of classification. The path from Locke to Buffon thus traced the first stages of a revolution in our approach to nature, from an approach marked by the search for divisions between the parts of nature to an attempt at something that could be equally attuned to its unity. Questions about generation, classification, natural history were defining investigations in the life sciences by the middle of the eighteenth century, and insofar as these formed the backdrop for Kant's own interests in natural history, the history of these questions must be examined as well.

Locke's approach to questions concerning the generation and classification of nature is best introduced by way of a brief reminder regarding Aristotle's and Boyle's roles in providing the backdrop for Locke's discussion. It is well understood that Aristotle's empirical investigations into organic processes were founded on his metaphysical account of the soul. Whether it was referred to as an animating principle or an entelechy, the soul explained the experience of a formative force in all living things; it made sense of life as an inner motion and of reproduction and growth as movement toward a specified goal.[23] In the seventeenth century, however, Aristotle's account of the souls of plants and animals was under attack from a number of fronts. The foremost of these attacks stemmed from religious precepts, flowing almost directly from Calvin's insistence that God's agency be accepted as the only source of activity in the natural world.[24] This position supported the kind of mechanical philosophy being promoted by Galileo and Descartes as well, since in their view nature was a realm filled with animate machines. From this philosophical perspective everything in nature was reducible to mechanical principles including, and especially, the organic body itself: the workings of muscle and tendon could be depicted as systems of pulleys, the heart likened to water bellows, and the nerves could be imagined to work like so many vibrating strings leading up to the head.[25] Calvin's Reformationist tenets thus easily combined with mechanical philosophy to describe nature as a collection of complex machines whose internal mechanisms were dependent upon God. But the central problem with this portrait of nature, a problem increasingly felt over the course of the seventeenth century, was that even the most elaborately imagined mechanisms could not account for the most constant experiences of organic life. They failed to explain the processes by which organisms were able to maintain and reproduce themselves, and they made no sense at all of the processes of inheritance despite the fact that breeders and horticulturalists were everywhere engaged in the attempted manipulation of them. And these sorts of everyday tensions between theory and practice were only compounded by the epistemic problems seen to be facing classification.

Because classification requires criteria for sorting, the determination of what can serve as criteria for this sorting is the first task in setting up a taxonomical system. For most of the history of classification leading up to Locke, the goal of taxonomy had been to create what systematists described as a "natural system," that is, a system that was capable of mirroring the divisions that were thought to exist within nature itself. The theoretical basis for this belief in natural divisions had been provided by Aristotle. In Aristotle's account, the formative force of the soul was responsible for directing organic processes toward a specified end, for moving an organism from a merely potential existence to a complete form. But in its formative capacity the soul not only explained, for example, why acorns become oaks; it was thought to serve also as the discriminating judge when it came to determining the essential features required for an oak to be an oak. It was as a result of this kind of work that nature could be understood to have divided itself up according to essential features, to have produced, in other words, a set of essential divisions underlying the possibility of a natural system.[26] But while Aristotle took such essential divisions to be real in nature, he was himself unconfident that the classificatory process of logical subordination could be adequately applied to biological life, for, as he saw it, it could never be clear to the taxonomist what nature itself had taken to be the essential or subordinate features of a given organism.[27] As Aristotle conceived of the problems facing taxonomy, the difficulties lay primarily on the side of the taxonomists and their ignorance with respect to nature's essential divisions. This problem went a step further for seventeenth-century mechanists, however, insofar as corpuscular ontology had rejected not only the soul as a basis for discerning essential differences between living organisms but the very notion of essential divisions existing within matter at all.

Corpuscular ontology had received its most concerted defense in the work of Robert Boyle, a thinker who was as much concerned with an extirpation of the chemical principles of Renaissance naturalism as he was with advancing his new corpuscular philosophy. He embraced corpuscular ontology in part, therefore, because it eliminated the possibility of irreducible elements—the mercury, salt, and sulfur of the Paracelsians—by taking matter to be substantially identical in all its parts.[28] Differentiation within matter, according to Boyle, occurred only as a result of shifts in the relative size, texture, and motion of the corpuscles. This meant that all material objects were the result of nonessential patterns of aggregation, patterns that had been produced by what Boyle described as a material "convention" or "stamp" upon an indifferent collection of matter.[29] But while this kind of corpuscular ontology allowed Boyle to respond to the iatrochemists, it also meant that he would be incapable of providing essential criteria by which inorganic matter could be meaningfully identified and sorted.[30]

When it came to accounting for organic matter, Boyle had appealed to a physicalist view of seminal principles. For Boyle, the sheer complexity of organic life exceeded the chance that its original formation had been due to the principles of secondary motion alone. Against the theory proposed by Descartes and his followers, therefore, Boyle argued for an original act of divine artifice that "did more particularly contrive some portions of that matter into seminal rudiments or principles, lodged in convenient receptacles (and, as it were, wombs), and others into the bodies of plants and animals." These seminal principles took on a formative function in directing the material unity of the organism, for "some juicy and spirituous parts of these living creatures must be fit to be turned into prolific seeds, whereby they might have a power, by generating their like, to propagate their species."[31] Although Boyle did not describe the exact means by which the formative work of the seminal principles operated, he clearly considered the process to be physical as opposed to soul driven:

I very well forsee it may be objected, that the Chick with all its parts is not a Mechanically contriv'd Engine, but fashion'd out of Matter by the Soul of the Bird . . . which by its Plastick power fashions the obsequious Matter, and becomes the Architect of its own Mansion. But not here to examine whether any animal, except Man, be other than a curious engine, I answer that this Objection invalidates not what I intend to prove from the alledg'd Example. For let the Plastick Principle be what it will, yet still, being a Physical Agent, it must act after a Physical manner, and having no other Matter to work upon but the White of the Egg, it can work upon that Matter but as Physical Agents, and consequently can but divide the Matter into minute parts of several Sizes and Shapes, and by local Motion variously context them.[32]

Boyle's commitment to a material interpretation of the work done by the seminal or plastic principle was clear from his appeals "Physical Agents."[33] Finishing the point, he explained "that the Formative Power (whatever that be) doth any more than guide these Motions, and thereby associate the fitted Particles of Matter after the manner requisite to constitute a Chick, is that which I think will not easily be evinc'd."[34]

Boyle's efforts to blend a corpuscular ontology with an account of seminal principles left open questions, however, regarding the coherence of mechanical approaches to nature. This incoherence was clearest with respect to taxonomical issues, since the ontology underlying the corpuscular theory of matter appeared to make classification impossible at the same time that the uneasy addition of materially conceived seminal principles were supposed to allow for it in the case of organic life. It was these strands in Boyle's thought that were most carefully taken up for consideration by John Locke. And it was here that Locke's own experience in medicine and botany would lead him to recognize the need to separate the problem of classification from the account of ontology. Taxonomy was a process of naming, according to Locke, and as such it was an endeavor that said more about decisions made by the taxonomist than it did about nature. And nothing could demonstrate the arbitrary nature of classification as much as could the fluid processes of organic generation and growth.

Locke's attitude toward the problems posed by biological generation developed in stages, with the first dating from his years at Oxford. As this time is well documented, it is perhaps enough here to recall that it was during these years that Locke learned of Descartes's mechanical philosophy; took a course on chemistry from the German Peter Stahl; read medical works by Harvey, Sennert, and the Galenists; created a personal *Herbarium*; and, of course, became acquainted with Robert Boyle and his corpuscular science.[35] It is in the so-called Morbus entry of 1666–1667, a text written while Locke was known to have been reading Boyle's *Origin of Forms and Qualities*, that we find an early response to the physical rendering of the "plastic principle" at work in generation. In this short and unfinished set of remarks, Locke was interested in determining "a more rational theory of diseases" based on the notion of seminal principles. As he defined them, "By seminal principles or ferments I mean some small and subtle parcels of matter which are apt to transmute far greater portions of matter into a new nature and new qualities."[36] Such principles, according to Locke, could perhaps explain the functioning of diseases, since these too seemed to transform the body's material into something new—that is, into the disease itself. Locke admitted that "how these small and insensible ferments, this potent archeus works I confess I cannot satisfactorily comprehend," but he was clear that it could not be operating according to the mechanical procedures that had been suggested by Boyle for the "straining" of particles by variously sized pores. As Locke saw it, only the transformative force of seminal principles could adequately explain the appearance of the "hard and consistent parts of the chicken" from out of the "soft and liquid" parts of the egg, and with respect to botany, only seminal principles could make sense of plant generation at all.[37] Describing this transformative force, Locke noted that "this change seems wholly to depend upon the operation or activity of this seminal principle, and not on the difference of the matter itself that is changed, so several seeds set in the same plot of earth change the moisture of the earth which is the common nourishment of them all into far different plants which differ both in their qualities and effects, which I think is not done by bare straining the nourishment through their pores which in different plants are of different shapes and sizes."[38] Regardless of how one is to interpret Locke's understanding of this "potent archeus" at work as the transformative force in generation, what the "Morbus" entry on disease makes clear above all is Locke's early skepticism regarding a mechanically reductive explanation of generation. This early hesitation can in fact be seen to have continued throughout Locke's work, even as his theories increasingly showed the influence of corpuscular science.

In 1667 Locke left Oxford for London, where he became for many years a close associate of Thomas Sydenham. Sydenham, typically described as England's foremost physician of the seventeenth century, was also interested in the problem of disease, and his widely read *Observationes Medicae* attempted to provide a natural history of the various species of disease on the models provided by botanical systems of classification. Like Locke, Sydenham took diseases to function by virtue of some kind of transformative power, a capacity to change the body's humors through the processes of "metamorphosis" into the disease itself. "The said humours," as Sydenham explained it, "become exalted into a *substantial form* or *species*; and these substantial forms or species manifest themselves in disorders coincident with their respective essences."[39] Sydenham's examples of this process of "exaltation" were always botanical, with mistletoe, moss, and fungi frequently cited as examples of a tree's essence having been transformed into a wholly new species.[40] Sydenham believed that a natural system could be created on the basis of essential features in the plant kingdom, and he took his investigations into the various courses taken by diseases to represent a parallel attempt. In his view, a natural history of diseases on this model would be invaluable, for it could form the backbone of a treatment program once diseases were definitively recognizable.

The preface to Sydenham's *Observationes Medicae* is considered to have been written either entirely by Locke or at least in close collaboration with him.[41] But given that the preface was published in 1676, when

Locke was already at work on drafts of the *Essay Concerning Human Understanding*, it seems clear that Sydenham's attempt to determine essential characteristics of disease would already have been at odds with Locke's emerging position on classification.[42] In a letter to Thomas Molyneux written after the publication of the *Essay*, for example, Locke was careful to distinguish the heuristic virtues of Sydenham's project—it could serve as an "art of memory" for the physician—from the possibility that such a thing could actually offer "philosophical truths to a naturalist." As Locke developed the point,

> Upon such Grounds as are the establish'd History of Diseases, *Hypotheses* might with less Danger be erected, which I think are so far useful, as they serve as an Art of Memory to direct the Physician in particular Cases, but not to be rely'd on as Foundations of Reasonings, or Verities to be contended for; they being, I think I may say of all of them, Suppositions taken up *gratis*, and will so remain, till we can discover how the natural Functions of the Body are performed, and by what Attraction of the Humours or Defects in the Parts they are hinder'd or disorder'd. . . . What we know of the works of Nature, especially in the Constitution of Health, and the Operation of our own Bodies, is only by the sensible Effects, but not by any certainty we can have of the Tools she uses or the Ways she works by.[43]

Locke's views here reflected the results of his discussion of taxonomy in the *Essay*, but before turning to the grounds he had provided for this position, it is worth recalling a few points regarding what we know of Locke's account of organic processes apart from the already cited comments made in his Morbus entry.

Like Boyle, Locke accepted seminal principles as at least a partial explanation for the original generation of both organic and nonorganic species. As he put it in his *Elements of Natural Philosophy* (1698), "All stones, metals, and minerals are real vegetables; that is, grow organically from proper seeds, as well as in plants."[44] Given his medical training, Locke was also familiar with theories that did not rely on seminal principles when explaining generation: the mechanical account on the model of fermentation provided by Descartes, the epigenetic version offered up by Harvey, and the preexistence theories taken to be supported by Anton van Leeuwenhoek's discovery of spermatozoa in 1677. Among the competing theories of generation, preexistence theorists argued that God had produced every single organic life form at the moment of creation. Depending upon the strain of preexistence theory, the individual life forms were then said to have been either embedded in the crust of the earth until they were taken up with food or to have been encased—the so-called Russian doll model—within either the ovaries or testes. But wherever these individuals were located after creation, they existed as submicroscopic yet fully formed organisms, and the gestation of an embryo was thus really only a process of mechanical enlargement. Although there would be problems for the theory in the long run, in their first appearances preexistence theories had a large number of supporters insofar as they fit with the mechanical approach to nature. It was this theory, for example, that lay at the heart of Locke's exchanges with Stillingfleet regarding resurrection. Locke was skeptical regarding the account, above all because it seemed impossible to assert anything like a material identity between a submicroscopic individual and a grown man.[45] His own view was that organic generation consisted in the rearrangement of previously created particles:

When a thing is made up of Particles, which did all of them before exist, but that very thing, so constituted of pre-existing Particles, which considered altogether make up such a Collection of simple *Ideas*, had not any *Existence* before, as this Man, this Egg, Rose, or Cherry, etc. And this, when referred to a Substance, produced in the ordinary course of Nature, by an internal Principle, but set on work by, and received from some external Agent, or Cause, and working by insensible ways, which we perceive not, we call *Generation*. (2.26.2)

Generation thus described the process by which an unsorted aggregate of preexisting particles was organized into a specific existence, into "this Man, this Egg"; how generation or the rearrangement of particles took place once the internal principle became active, however, was something Locke considered to be incomprehensible.

Locke was also familiar with botanical processes, for he had actively built up a collection of plants for his own *Herbarium*—a catalog remaining one of the best preserved from that century—taking careful note of species, hybrids, and random mutations such as a blue flower appearing among the expected yellow. Compared to the general constancy of animal reproduction, Locke thus noted at one point that "in vegetables we find that several sorts come from the seeds of one and the same individual as much different species as are allowed to be so by the philosophers."[46] And he worked to keep abreast of the ongoing changes and debates in botany regarding the classification of particular species of plants during this period, noting changes that had affected his own catalog and meeting with horticulturalists to discuss the results.[47] Locke's early engagement with the problem of understanding natural processes—whether regarding the transformative power of disease or the internal principle at work in generation—would combine to support his views regarding classification. In particular, it seems to have convinced him that classification should disentangle itself as much as possible from any kind of ontological commitments regarding the things being classified.

As explained above, the main theoretical task facing classification practices in the seventeenth century was determining the criteria that would be used for sorting whatever objects were under consideration. Once this theoretical task had been accomplished, then it was supposed to be only a practical matter with respect to sorting these individuals into groups according to the criteria that had been set. With respect to the theoretical task, the guiding assumption was that the system of nature could only be understood—and thereby classified—if nature's own taxonomical criteria could be discovered. Such discovery however, as taxonomists widely recognized, presented an almost insuperable challenge. Adding to the theoretical challenge facing taxonomists was the practical problem of having to deal with organisms—and plants were particularly difficult in this way—that seemed resolutely indeterminate, that is, that showed characteristics placing them in two or even three separate categories at once. Locke understood that these were the central difficulties facing natural history, but he also thought that these problems had mainly to do with the incorrectly perceived terms under which taxonomists were laboring. It was not obvious to him that nature should even be interested in maintaining boundaries between species, nor was it clear, with all the shape-shifting going on in the plant world, for example, that such boundaries could ever be meaningfully maintained. The natural system, as Locke saw it, was an unsupportable myth, and the sooner taxonomists recognized this fact, the more likely it was that classification might make some progress toward an adequate system.

Classification was a human practice meant for human ends, and the problem facing classification thus lay in a separate direction altogether, since it was essentially tied to facts about cognition. All sorting was the "Workmanship of the Understanding" (3.3.12) for Locke, and as such it was open to the vagaries of individual judgment as well; as he put it, it *"depends upon the various Care, Industry, or Fancy of him that makes it"* (3.6.29). For example, "if the *Idea* of *Body* be bare Extension or Space," according to one person, "then Solidity is not *essential* to Body: If others make the *Idea*, to which they give the name *Body*, to be Solidity and Extension, then Solidity is essential to *Body*. That therefore, and *that alone is considered as essential, which makes a part of the complex* Idea *the name of a sort stands for,*" according to Locke, and in this sense, "to talk of specifick Differences in Nature, without reference to general *Ideas* and Names, is to talk unintelligibly" (3.6.5). It was therefore the naming of things, or rather the annexing of a name to a particular abstract idea that one had formed, that alone determined species. The supposed real essence of a determined kind was ultimately unknowable, even in the case of mankind, and Locke pointed to comas, delirium, retardation, and madness, all in the effort to undermine any sense that rationality might prove to be an exception to this fact (3.6.29).[48]

Because classification was driven by pragmatic considerations regarding communication and order, it did not make sense to assume that nature could be similarly invested in determining boundaries between species. As Locke made the point,

> Wherein then, would I gladly know, consists the precise and *unmovable Boundaries* of that *Species*? 'Tis plain, if we examine, there is *no* such thing *made by Nature,* and established by Her amongst Men. . . . So uncertain are the Boundaries of *Species* of Animals to us, who have no other Measures, than the complex *Ideas* of our own collecting: And so far are we from certainly knowing what a Man is; though, perhaps, it will be judged great Ignorance to make any doubt about it. And yet, I think, I may say, that the certain Boundaries of the *Species*, are so far from being determined, and the precise number of simple *Ideas* which make the nominal essence so far from settled, and perfectly known, that very material Doubts may still arise from it. (3.6.27)

It was in fact the "very material doubts" arising from attempts to determine natural kinds that indicated at once not only the artificial nature of our classification system but the actual imprecision of nature itself. In keeping with this, Locke repeatedly offered examples of hybrids, deformation, and even mythical creatures to make the point regarding both nature's plasticity and the impossibility that independently established categories could ever make sense of that fluidity.[49] "Nor let anyone say," as he put it, "that the power of propagation in animals by the mixture of Male and Female, and in Plants by Seeds, keeps the supposed real *Species* distinct and entire . . . for if history lie not, Women have conceived by Drills; and what real *Species*, by that measure, such a Production will be in Nature, will be a new question" (3.6.23). It was with respect to this natural fluidity that Locke resorted to the role played by "life," moreover, when it came to understanding organic unity at all.

As he described it, organic unity was maintained only insofar as the organization of parts could be collectively orchestrated by their partaking in a common life. "That being then one Plant," he explained,

> which has such an Organization of Parts in one coherent Body, partaking of one Common Life, that it continues to be the same Plant, as long as it partakes of the same Life, though that Life be communicated to new Particles of Matter vitally united to the living Plant in a like continued Organisation, conformable to that sort of plants. For this Organisation being at any one instant in any one collection of *Matter*, is in that particular concrete distinguished from all others, and is that individual Life, which existing constantly from that moment both forwards and backwards in the same continuity of insensibly succeeding Parts united to the living Body of the Plant, it has that Identity, which makes the same Plant, and all the parts of it, parts of the same Plant, during all the time that they exist united in that continued Organisation, which is fit to convey that Common Life to all the Parts so united. (2.27.4)

The concept of life served thus as a constantly unifying force within the "insensibly succeeding Parts" of the plant (2.27.4). Life was more than the organism's "collection of matter," because it was the active principle generating an individual life, an identity so long as the parts were orchestrated together by it.[50]

But while Locke seems to have both respected the general irreducibility of organic processes and demanded that classification be recognized as something that was entirely the "workmanship of the understanding," he was insistent that our ideas of substances stood independent of such complete workmanship. It was precisely because the "patterns" of our ideas of substances lay outside us, according to Locke, that we could not achieve the level of certainty and coherence afforded either mathematics or our ideas of morality, religion, and politics. In these modes of thinking, the patterns or "archetypes" lay within the mind itself (4.1.1); in the case of substances, our ideas were in some sense original to the substance itself. And it was in this vein—that is, in the distinction between substances understood to be really existing outside of us and ideas that do not—that Locke took it to be a matter of common sense for us to assume real differences in the "internal constitution" of things (e.g., 3.6.6, 3.6.9, 3.6.28), particularly as this fit with his belief that such a "real essence" bore a causal relationship to our sensible ideas.[51] For Locke, the reality of individuals was simply both a given and distinct from arguments regarding the logic of classification.[52] As he wrote to William Molyneux,

In the objection you raise about species I fear you are fallen into the same difficulty I often found my self under when I was writing of that subject, where I was very apt to suppose distinct species I could talk of without names. For pray, Sir, consider what it is you mean when you say, that *we can no more doubt of a sparrow's being a bird, and an horse's being a beast, than we can of this colour being black, and t'other white,* etc. but this, that the combination of simple ideas which the word bird stands for, is to be found in that particular thing we call a sparrow. And therefore I hope I have no where said, *there is no such sort of creatures in nature as birds*; if I have, it is both contrary to truth and to my opinion. This I do say, that there are real constitutions in things from whence these simple ideas flow, which we observ'd combined in them. And this I farther say, that there are real distinctions and differences in those real constitutions one from another; whereby they are distinguished one from another, whether we think of them or name them or no. But that that whereby we distinguish and rank particular substances into sorts or genera and species, are not those real essences or internal constitutions, but such combinations of simple ideas as we observe in them.[53]

For Locke, then, "there are things from whence ideas flow," and there are "real distinctions and differences in those real constitutions," but these were not in any sense to be understood as providing the criteria for their subsequent sorting. Real essence could not be known, according to Locke, though its effects—the existence of its external "pattern"—could be somehow recognized when receiving material sensations. Thus, while it was a matter of common sense to assume real differences between substances, this fact in no way influenced Locke's conclusions regarding the actual process by which classification occurred: "'Tis true, I have often mentioned a *real Essence*, distinct in Substances, from those abstract *Ideas* of them, which I call their *nominal Essence*.... But [real] *Essence*, even in this sense, *relates to a Sort*, and supposes a *Species*: for being that real Constitution, on which the Properties depend, it necessarily supposes a sort of Things, Properties belonging only to *Species*, and not to Individuals ... [for] there is no individual parcel of matter, to which any of these Qualities are so annexed, as to be *essential* to it, or inseparable from it" (3.6.6).[54] Locke's species nominalism did not entail a lack of commitment on his part to the real existence of individual substances, therefore, but this commitment did not itself mean that Locke would ever agree that essential features could somehow be logically determined in the absence of criteria for sorting.[55] Locke was both a nominalist regarding species determination and a realist in believing that there were inner features contributing to species as well. In a similar fashion, Locke was both comfortable with a mechanical portrait of animal functioning and cognizant of the need for "inner principles" and "transformative forces" when it came to understanding the processes of organic life. And all of this contributed to Locke's views of both nature and the proper task of classification.

Reviewing Locke's early considerations of organic processes against the backdrop of corpuscular ontology reveals his sensitivity to the problems facing Boyle in the case of organic life. While Locke remained committed to the essential features of corpuscular science, he was nonetheless hesitant in the face of a straightforward endorsement of mechanical accounts of generation. For the problem with that approach, as Locke summarized it in *Some Thoughts Concerning Education*, was that it "leaves no room for the Admittance of Spirits, or the allowing any such things as *immaterial Beings in rerum natura*: when yet it is evident, that by mere Matter and Motion, none of the great Phænomena of Nature can be resolved."[56]

Leibniz's Organic Machines

While Locke might have been suspicious of mechanical accounts of organic generation, he completely rejected an increasingly popular attempt to save mechanical principles, namely, the preexistence theory of generation. Appeals to God's original production of both matter and the seminal principles explaining the origin of species were common throughout the seventeenth century, and Boyle and Locke were mainstream in endorsing this approach. There had been rising dissatisfaction with the effort to describe individual generation by way of mechanics, however—Descartes's fermentation as the site of inner force or Boyle's plastic principle as only a motion fitting together the parts, for example—and this had encouraged renewed interest in the conceptual possibilities afforded by seminal principles as means for thinking about individual generation as well. In this manner preexistence theories of generation proposed that individuals could be thought of along the lines of submicroscopic seeds, seeds whose generation had occurred at creation. Mechanical principles found a place in this account, since the gradual enlargement or expansion of these miniscule preformed individuals was easy to imagine through mechanical models of pumps and vacuums. The conceptual advantages of this position were obvious: as God was already taken to be the author of all creation, it was hardly a stretch to suppose that God had in fact created all future generations of organic life in that first act as well. This fit well, moreover, with Calvinist tenets regarding the passivity of matter. Thus while details varied between theorists regarding the actual process by which these generations of individuals transitioned from preexisting seed to developed organism, on the whole, the tasks associated with this description were conceptually preferable to describing the formation of individuals by means of motion alone. What is more, Leeuwenhoek's discovery of spermatozoa appeared to promise physical evidence that the general theory might be true.[57]

Leeuwenhoek's investigations, along with those of the other two important microscopists, Jan Swammerdam and Marcello Malpighi, were taken to be especially significant by Gottfried Leibniz. For Leibniz, Leeuwenhoek's 1674 discovery of life teeming in a drop of pond water appeared to provide empirical support for a metaphysical system that was meant to challenge the view of nature supported by Locke and Newton.[58] The various discoveries being made by the microscopists had been the subject of much discussion while Leibniz was in Paris between 1672 and 1676, and his return to Hanover was preceded by a trip to Holland, where he met with both Swammerdam and Leeuwenhoek. Leibniz's ultimate view was that individuals were composed of living monads arranged hierarchically under a dominant entelechy or soul.[59] As he summarized this in the *Monadology*,

From this we can see that there is a world of creatures, of living beings, of animals, of entelechies, of souls in the least part of matter. Each portion of matter can be conceived as a garden full of plants, and as a pond full of fish. But each branch of a plant, each limb of an animal, each drop of its humours, is still another such garden or pond. . . . Thus we see that each living body has a dominant entelechy, which in the animal is the soul; but the limbs of this living body are full of other living beings, plants, animals, each of which also has its entelechy, or its dominant soul. . . . But we must not imagine, as some who have misunderstood my thought do, that each soul has a mass or portion of matter of its own, always proper to or allotted by it, and that it consequently possesses other lower living beings, forever destined to serve it. For all bodies are in a perpetual flux, like rivers, and parts enter into them and depart from them continually.[60]

This position was reached after numerous considerations, not the least of which concerned the problem facing all corpuscular accounts regarding material unity. Unity, Leibniz concluded, could only be the result of an organizing force, for it is impossible to find *the principles of a true unity* in matter alone, or in what is only passive, since everything in it is only a collection or aggregation of parts to infinity.... Hence it was necessary to restore, and, as it were to rehabilitate the *substantial forms* which are in such disrepute today but in a way that would render them intelligible.... Aristotle calls them *first entelechies;* I call them, perhaps more intelligibly, *primitive forces,* which contain not only *act* or the completion of possibility, but also an original *activity.*[61]

These primitive active forces explained the metaphysical possibility of unity both for the individual monad and the organic whole, a whole organized as such so far as the dominant monad or entelechy determined the subordinate monads to a specific end.

Like Aristotle's soul, Leibniz's conception took primitive active force to be the only explanation for both the form of a substance and the force required to achieve it. In Leibniz's scheme, however, corporeal substance teemed with life, and it did so in such a manner that physical changes could be understood as the gradual shifting in dominance from one monad to another. What the microscopists provided, therefore, was empirical support for the metaphysical dimension of Leibniz's theory insofar as their discoveries pointed to a continuum of life, a continuum undergirded, in Leibniz's view, by the repeated transformations of corporeal substance.[62] Leibniz was thus happy to report, for example, that "on the basis of very important analogies in anatomy, Mr. Malpighi is strongly inclined to believe that plants can be included in the same genus with animals and that they are imperfect animals."[63] Appealing to Swammerdam, Leibniz used the supposed anatomical similarity between organs of respiration in plants and animals as another example of this continuum. "Mr. Swammerdam has supplied observations which show that insects are close to plants with respect to their organs of respiration and that there is a definite order of descent in nature from animals to plants."[64] And regarding Leeuwenhoek's infusoria, Leibniz wrote to Antoine Arnauld, "Those who conceive that there is as it were an infinity of small animals in the least drop of water, as Mr. Leeuwenhoek has shown, and who do not find it strange that matter should be filled everywhere with animated substances, will not find it any more strange that there is something animated even in ashes, so that fire can transform an animal and reduce it to small size, instead of destroying it entirely. What can be said of one caterpillar or silkworm can be said of a hundred or a thousand animals."[65]

A number of consequences flowed directly from this view, including the sense that matter and soul were inseparable and that there could never be, therefore, a transmigration or "metempsychosis" of the soul apart from the body. If the soul was indestructible, then the body must be too.[66] Leibniz thus described the appearance of dramatic physical changes—changes resulting from the apparent generation or corruption of individuals—as in fact only transformations, an *augmentation* or *diminution* of the organic machine.[67] Once more, the microscopists were called upon in support of Leibniz's position so far as their work, and Leeuwenhoek's 1677 discovery of spermatic animalcules in particular, seemed to provide physical evidence of such diminutive individuals.[68] As Leibniz put it,

This is where the *transformations* of Swammerdam, Malpighi, and Leeuwenhoek, the best observers of our time, have come to my aid and have made it easier for me to admit that animals and all other organized substances have no beginning, although we think they do, and that their apparent generation is only a development, a kind of augmentation.

It is therefore natural that an animal, having always been alive and organized (as some persons of great insight are beginning to recognize), always remain so. And since there is no first birth or entirely new generation of an animal, it follows that there will not be any final extinction or complete death, in a metaphysical sense. Consequently, instead of the *transmigration* of souls, there is only a *transformation* of the same animal, according to whether its organs are differently enfolded and more or less developed.[69]

Despite Leibniz's appeals to the discoveries of the microscopists, however, it must be remembered that Leibniz in no way considered his philosophical account to be dependent upon empirical evidence produced by the life sciences. Indeed he wavered between ovism and animalculism without any sense that one version might support his position better than another. As he wrote in a late letter to his follower Louis Bourguet,

I very much wish that we could go further into the great issue of the generation of animals, which must have an analogy with that of plants. Mr. Camerarius of Tübingen thought that their seed is like the ovary, and the pollen (although in the same plant) like the sperm of the male. But even if that were true, the question would always remain whether the basis of the transformation, or the preformed living thing, is in the ovary, following Mr. Vallisinieri, or in the sperm, following Mr. Leeuwenhoek. For I hold that there must always be a preformed living thing, whether plant or animal, which is the basis of the transformation, and that the same dominant monad be in it.[70]

CHAPTER ONE

Leibniz had followed corpuscularian philosophy in holding that God had created all matter and could alone bring about its destruction. What preexistence theories of generation argued for, and what the results of the microscopists seemed to support, was that this was true for organic material as well. As Leibniz understood it, not only was this latter point right, but it also meant that all individuals were the result of their having been formed by God at the origin of the world.[71]

In the same manner that all seeds or individuals had been originally created, there was, according to Leibniz, a divine preformation at work when it came to ideas as well. "The mathematical sciences," Leibniz explained, "which deal with eternal truths rooted in the divine mind, prepare us for the knowledge of substances," so that although very little can actually be known with this kind of distinctness, "the seeds of the things we learn are within us—the ideas and the eternal truths which arise from them."[72] For Leibniz it was the affinity or shared origin of the preformed mind and its preformed ideas that grounded the necessity ascribed to truths of reason. In his words,

What makes the exercise of the faculty easy and natural so far as these truths are concerned is a special affinity which the human mind has with them; and that is what makes us call them innate. So it is not a bare faculty, consisting in a mere possibility of understanding those truths: it is rather a disposition, an aptitude, a preformation, which determines our soul and brings it about that they are derivable from it.[73]

When Leeuwenhoek had discovered the parthenogenesis or "virgin birth" of aphids in 1694, he was able to use his discovery as a model for understanding the seminal production of animalcules.[74] From Leibniz's perspective, the mode of virgin birth performed by the aphids neatly mirrored his account of the "virgin" generation of truths from seeds that had been implanted by God.

Leibniz's emphasis on the special affinity between mind and idea argued directly against empiricist tenets regarding the origin of knowledge so far as he took necessary truths to be the realization of a predetermined disposition on the occasion of experience. At the same time, it was precisely this special affinity or shared origin of idea, truth, and the apperceptive monad that guaranteed necessity to the truths of reason. For according to Leibniz, there was a difference between contingently discovered and necessary truths, "just as there is a difference," as he put it, "between the shapes which are arbitrarily given to a stone or piece of marble, and those which its veins already indicate or are disposed to indicate if the sculptor avails himself of them."[75] This was why Leibniz ultimately believed that "the innate concepts of Plato, which he concealed by the term 'reminiscence,' are therefore by far to be preferred to the blank tablets of Aristotle, Locke, and other exoteric philosophers," for Plato too understood the significance of a nonempirical origin when it came to establishing truth.[76]

Leibniz's position, borrowing as it had from Plato and Aristotle as much as from the life sciences and theology, ultimately found itself under attack, particularly by Newtonian partisans in the wake of the Leibniz-Clark controversy. It was an attack that found its initial focus on the topic of force. But while much has been written on the specific arguments at work in the *vis viva* debate, what is important for the purposes of this discussion is to see how the debate over force eventually contributed to a turning point in discussions of biological generation. The preexistence theory of generation—in all of its forms, from ovism to spermatic animalcules to Perrault's germ theory—was successful so far as at filled an explanatory gap left by the mechanical philosophy of Descartes and Boyle. Leibniz's own support for divine preformation, however, hardly had to do with an interest on his part in preserving Cartesian views of nature. On the contrary, divine preformation was maintained by Leibniz on grounds that had primarily to do with his metaphysics. When discussion turned to forces, Leibniz approached problems in mechanics in precisely the same manner, taking physical forces to be likewise grounded by a metaphysical account. In the *vis viva* debate, for example, Leibniz had argued that quantity of motion could not serve as an adequate measure of force, substituting in its place an active force (mv^2) that was ultimately demonstrated to be correct by the Newtonian Willem 'sGravesande.[77] But this active force concerned only the derivative forces of the gross bodies of physics. And derivative forces, whether active (*vis activa*) or passive (*vis mortua*), could only be meaningful so far as they were grounded on the doctrine of the monads, specifically the primitive active force Leibniz referred to as the soul or entelechy.[78] It was this move, the grounding of derivative forces on a metaphysical basis, that meant that Leibniz's mechanics as much as his endorsement of a preexistence theory of generation, would be swept away in favor of the greater promise now felt to attend the possibilities of a purely materialist account, one newly energized by Newton's discussion of attractive and repulsive material forces.

The Newtonians would decry Leibniz's notion of an inner force as a return to Aristotle and Scholastic metaphysics, and Leibniz would declare in turn that the application of active attractive and repulsive forces at the level of physics could only amount to a revival of the occult forces of Renaissance naturalism and thereby the end of true mechanism. Indeed, the explanatory power of mechanical principles could only be maintained, as Leibniz saw it, so long as the operating causes of metaphysics and physics remained distinct.[79] Leibniz might have had 'sGravesande's empirical demonstration on his side when it came to the question of active force, but Newtonianism was in its ascendency, and the result of the *vis viva* controversy was the sense that forces that had performed so admirably in the service of mechanics might just as well be adapted to ends in the life sciences. What is more, as pressures mounted against preexistence theories, an avenue seemed to have opened up for the rehabilitation of a mechanical theory of generation, one powered by material forces and thus no longer reliant on metaphysical conceptions of entelechies or soul.

TWO

Buffon's Natural History and the Founding of Organicism

Hales and the Physiology of Plants

Few concepts in eighteenth-century science would prove to be as plastic as the concept of "force." Newton's *Optics* had become famous as much for its "Queries" as for its account of light, and Newton's suggestion there that the same understanding of the forces at work in physics might be applied to chemistry would turn into a research program for much of the century to come. Until the 1740s, the majority of researchers assumed a continuum between the inorganic and organic realms, and this continuum ensured that the language of forces would be applied to organic bodies as well. This began very much in the shadow of Newton with Stephen Hales's mechanical view of plant physiology, and this strain would continue in Albrecht von Haller's identification of the "sensible" and "irritable" forces at work in human physiology.[80] The mechanical model dominated initial applications of forces at work in theories of organic generation as well. In 1729 Leibniz's disciple Louis Bourguet introduced the language of "organic mechanism" to distinguish the necessary interiority of organic growth—Bourguet followed René Réaumur in calling this kind of growth "intussusception"—from the kind of external accumulation at work in crystal formation, an account that Buffon would build upon with the addition of a "penetrating force" to guide the organic process.[81] By the 1780s "organic forces" or "emergent vital forces," like Caspar Wolff's *vis essentialis* and Johann Blumenbach's *Bildungstrieb*, would come to dominate the life sciences, shedding at last the demand for exclusively mechanical models of nature. The crucial factor for the establishment of such organicism, however, was the slow establishment of natural history as a science oriented toward the temporal histories of species, and for that, naturalists could thank Georges Buffon above all.

Buffon's role, in this narrative at least, begins with the work of Stephen Hales. One of the first and in some sense most influential applications of Newtonian forces to an organic system had been presented to the Royal Society in 1727 by Hales as a set of statistical results or "Vegetable Staticks," generated by the experiments Hales would go on to enumerate.[82] Hales was not primarily an anatomist, nor was he at all interested in the problems of taxonomy. What Hales wanted to address was the set of questions surrounding "plant physics," or the physiology of plants: the movement of sap, the management of temperature, the processes of nutrition, and above all, the relationship between plants and air. The result was some two hundred pages of carefully described experiments, many including plates meant to show a particular configuration or a special apparatus designed for some test, and an overall reticence when it came to speculation regarding specific results. This, combined with Hales's attention to the practical use of his results for problems in agriculture, made him a perfect representative of the ideals set out by the Royal Society at the time. Questions regarding circulation and respiration had indeed been of enduring interest in the society, and this was where Hales began his own investigations in 1718, looking for means to test the circulation of sap. Hales was mainstream in his assuming there to be essential parallels between the physiology of plants and animals, and his belief that "as in vegetables, so doubtless in animals" led his hypotheses for numerous experiments.[83] It was on the basis of this analogy that Hales was led from his investigations into circulation to the question of respiration—experiments had revealed that blood turned a lighter color after passing through the lungs of animals—and Hales quickly decided that leaves were the "lungs of the plant."[84] His experiments on respiration were considered to be the most significant yet performed toward discovering the properties of air, and Hales applied his understanding of Newton's forces here above all. According to Hales, the attractive and repulsive forces at work in plant respiration operated by means of the attracting properties of "fixed air" in helping to form the solid parts of the plant (a fact evidenced by the amount of air released during fermentation) and the counteracting repulsive force of "elastic air" for the promotion of plant growth.[85]

Hales considered air—"this now fixt, now volatile *Proteus*"—to be key for understanding generation.[86] At this point fertilization was not yet fully understood, but Hales followed others in rightly taking the relationship between the "farina" of the anthers and the bulbous pistil to be critical for producing fertile seeds. The main question facing investigators at this stage concerned the seeming impossibility of any means for the pollen to reach seeds tightly encased within the pistil. Air, Hales now suggested, combined with the active principles of light and sulfur, might together form "a *Punctum Saliens* to invigorate the *seminal* plant" and thereby yield an "unhatched tree." [87] Hales described this "tree-egg" as follows: "As soon as the *Calix* is formed into a small fruit, now impregnated with its minute seminal tree . . . (which new set fruit may in that state be looked upon as a complete egg of the tree, containing its young unhatched tree, yet in embryo) then the blossom falls off, leaving this new formed egg, or first set fruit in this infant state, to imbibe nourishment for itself and the Foetus with which it is impregnated."[88] By the time Hales came to these conclusions, he would have been well familiar with the work of another important fellow of the Royal Society, Patrick Blair. The publishers for the Royal Society had put out Blair's influential botanical essays in 1720, and since then Blair had in part sought to combat animalculist theories of preexistence with his own evidence.[89] In 1721 Blair published in the society's *Philosophical Transactions* a letter detailing the findings of local gardeners regarding plant generation that included a discussion of the formation of hybrids, a phenomenon Blair took to be key evidence against Leeuwenhoek's experimental support for the preexistence theory of generation.[90] Hybrids, as Blair understood them, proved that the male and female parent each contributed materially to the formation of the embryo, "which could never happen did these organized *Animalcula*, or granules of the *Farina* become a *foetus*, or contain the *folia seminalia* of a plant."[91] Generation, according to Blair, took place once the pollen—a substance serving only as the vehicle for a vivifying spirit—landed in the flower cup and allowed the *"vivificke Effluvia"* to fertilize the seed. Hales's *Punctum saliens* clearly pursued this model for insemination as well.

While Hales and Blair were concerned to discover the processes of generation, they accepted the sexuality of plants as a matter of fact. The idea that the different organs of the flower (in the case of hermaphroditic plants) or different flowers of the same species could be understood on analogy with the sexual organs of animals was established in 1694 by Rudolf Jacob Camerarius.[92] Although Camerarius's letter on the subject had been initially published in the relatively obscure transactions of a Tübingen learned society, his results soon became widely known, so that by 1718 Herman Boerhaave, for example, was teaching Camerarius's ideas, and Sébastien Vaillant, a botanist at the Jardin du Roi in Paris, published a public lecture on the sexuality of plants that had been given to an audience of some six hundred at the Royal Gardens.[93] The impact of this idea for the life sciences of the time cannot be underestimated, for what it suggested was a potential identity between the laws of generation in the plant and animal kingdoms, and while research into animal physiology was both difficult and messy, the cultivation of plants—in some sense, the most plastic of all forms of organic life—could be open to anyone. Once Hales's application of Newtonian forces to the physiological processes of plants became known, therefore, it was only a matter of time before these would be applied to the animal kingdom as well.

Buffon the French Newtonian

The significance of Hales's research rapidly spread throughout Europe as a result, in part, of its having been translated into French.[94] Then as now, the translation of an important work could be the source of some prestige for the beginning academic, and Georges Buffon's decision to translate Hales thus anticipated what would be a career of successful strategizing on his part. That said, Buffon's choice to translate Hales also reflected deep affinities between the approach Hales took and those of the so-called French Newtonian. Buffon took his time with the translation, adding notes where he felt the explanations were lacking, toning down some of the more overtly deist sentiments, and augmenting when possible the Newtonian cast of descriptions.[95] The translation thus marked a period when Buffon's positivist attitude toward science was at its most pronounced. After extolling the virtues of Hales's work in his preface, for example, Buffon turned to what he took at that point to be the proper method in science:

The furnishings of nature may well rest on distinct parts and principles but their distinctness remains as unknown to us as their interconnection. Now how could one measure these secrets with only the imagination or some other such means of discovery; and how could we forget that we find nothing else before us than effects and that these alone should be the means for researching into their causes? Only through precise and correct reflection on constant experience have we compelled nature to reveal her secrets. . . . This is the method that my author has followed; it is that of the great Newton.[96]

Buffon had been attracted to Newton by the calculus, and after Hales his next translation project was the French edition of Newton's essay "The Method of Fluxions and Infinite Series"—a project chosen, as Buffon's preface suggests, in part for the opportunity it gave for defending the calculus against claims to authorship by the Leibnizians.[97]

Newton's influence on Buffon yielded results that would have wide-ranging consequences for the work that would be done in the latter's *Natural History*. One source of influence lay in Newton's use of mathematics when describing natural events. For his own part, Buffon had long been interested in the application of probability theory to games of chance, and his first publication—a piece written toward his admission to the French Academy of Sciences—attempted to combine the calculus and geometry in order to generate probable outcomes in a game of chance. This game, *franc-carreau*, asked bettors to guess how many cracks would be crossed were a tossed coin to land on a tiled floor. Here Buffon concentrated on the difference in outcomes given shifts in relative proportions; big coins on small tiles, in other words, would lend easy advantage to those betting on a higher number of cracks, so the mathematical problem was to determine the possibility of a fair distribution of chance for all the bettors. Buffon's attention to this problem convinced him that the work of calculating probable outcomes admitted ready application across the sciences.

Buffon therefore made probability theory an integral part of his methodology when approaching natural history, introducing his idea for the "true method" in natural history in the "Initial Discourse," the opening piece for the first three volumes of his *Natural History*, which appeared in 1749.[98] This method consisted in the synthesis of two kinds of truths, the mathematical and the physical. Buffon distinguished "physical truth," as the inductive, a posteriori gathering of "facts," from the arid results produced by "mathematical truths," truths bearing no tincture of the real. The two could be profitably combined, however, in a manner similar to the earlier application of probability theory to games of chance. In this case, the application of mathematics to the inductive and thus contingent results based on physics would yield conclusions that while merely probable were nonetheless "equivalent to certitude." As Buffon stated it,

> It is here that the union of the two sciences of mathematics and physics might result in great advantages. The one gives the "how many," the other the "how" of things. And since it is a question here of combining and estimating probabilities in order to judge whether an effect depends more on one cause than on another, when you have imagined by physics the how, that is to say, when you have seen that such and such an effect might well depend upon such and such a cause, you then apply mathematics in order to assure yourself as to how often this effect happens in conjunction with its cause. And if you find that the result accords with the observations, the probability that you have guessed correctly is so increased that it becomes a certainty.[99]

This synthesis of mathematical and physical truths or the "true method" for natural history allowed the investigator to capture nature from two perspectives at once. Nature could be viewed on the one hand as a unified system of species whose formal contours had been determined by God at creation—an eternal view of species—and as the successive series of temporally determined individuals on the other hand. To achieve this the naturalist needed to combine a talent for seeing individuals in all their specificity—their birth, generation, organization, and habits—while reflecting also on the long view of the history of a species as a whole, as something more than the mere aggregation of individuals. According to Buffon, "A vast memory, assiduity, and attention suffice to arrive at the first end. But more is needed here. General views, a steady eye, and a process of reasoning informed more by reflection than by study are what is called for. Finally, that quality of the mind is needed which makes us capable of grasping distant relationships, bringing them together, and making out of them a body of reasoned ideas after having precisely determined their nearness to truth and weighed their probabilities."[100] Despite the fact that Buffon's admittance to the French Academy of Sciences had been due to his early work on mathematical probabilities, he never ceased to insist upon the need for a "reflective synthesis" of mathematical and physical truths if there was to be any hope for advancing investigations into natural history. Against the Cartesian approach, "mathematical truths," as Buffon instead understood them, "are only exact repetitions of definitions"; they have "nothing of the real" and as merely "different expressions of the same thing" they have the advantage of always being "precise and conclusive, but abstract, intellectual, and arbitrary" and are therefore incapable of providing anything like a real portrait of nature. Physical truths, by contrast, concerned "facts": "A sequence of similar facts or, if you prefer, a frequent repetition and an uninterrupted succession of the same occurrences constitute the essence of this sort of truth. What is called physical truth is thus only a probability, but a probability so great that it is equivalent to certainty."[101] In order to achieve a genuine natural history, one whose claims were a synthesis of exacting description and historical explanation, both kinds of truth were needed.

The application of the true method to natural history was new, but the synthesis itself represented an approach that Buffon took to be already at work in Newton.[102] Newtonian forces served as a means for understanding a general unity in nature from the greatest cosmological relations to the minute workings of chemical affinity.[103] And following the model supplied by Newton via Hales, it was indeed through the workings of such "penetrating forces" that Buffon understood the physiological processes of generation to be operating. Buffon's descriptions of the "organic molecules"—variously described as "living matter" and "active principles"—upon whom the forces were at work, however, recalled nothing so much as Leibniz's well-known discussions from the *Monadology*. Like Leibniz, Buffon took these molecules to be the living matter originally determined by God to be the basis of physical existence. As Buffon described it,

God, when he created the first individuals of each species of animal and vegetable, not only bestowed form on the dust of the earth, but gave it animation, by infusing into these individuals a greater or smaller quantity of active principles, of living organic particles, which are indestructible and common to every organized being. These particles pass from body to body, and are equally causes of life, of the continuation of the species, of growth, and of nutrition. After the dissolution of the body, after it is reduced to ashes, these organic particles, upon which death has no influence, survive, circulate through the universe, pass into other beings, and produce life and nourishment. Hence, every production, every renovation or increase by means of generation, of nutrition, or of growth, implies a preceding destruction, a conversion of substance, a translation of organic particles, which never multiply, but, uniformly subsisting in equal numbers, render Nature always equally animated, the earth equally peopled, and equally resplendent with the original glory of that Being by whom it was created.[104]

Ingested as food, the organic particles were diffused throughout the body, allowing for its nutrition and growth. At puberty the body was fully grown, and the excessive particles returned to the sexual organs

CHAPTER TWO

bearing impressions of the body's internal "mold," an artifice produced for the "first individual of each species" by God but thereafter mechanically replicated by the actions of the molecules and a penetrating force.[105] "What can be the active power which causes this organic matter to penetrate and incorporate itself with this internal mould?"[106] For Buffon it was the penetrating force, a notion not only modeled on Newtonian forces but one that in its explanatory role paralleled the job assigned by Newton to gravity. Indeed Newton's appeal to gravity as an unknown source of nonetheless demonstrable effects offered an epistemic model for physiologists throughout the eighteenth century dealing with similar physiological unknowns when describing phenomena.[107] As Buffon explained the working of this force,

> In the same mode as gravity penetrates all parts of matter, so the power which impels or attracts the organic particles of food, penetrates into the internal parts of organized bodies, and as those bodies have a certain form, which we call the internal mould, the organic particles, impelled by the action of the penetrating force, cannot enter therein but in a certain order relative to this form, which consequently it cannot change, but only augment its dimensions, and thus produce the growth of organized bodies; and if in the organized body, expanded by these means, there are some particles whose external and internal forms are like that of the whole body, from those reproduction will proceed.[108]

For all the similarities between Buffon's organic molecules and Leibniz's monads, the result of Buffon's apparent borrowing from Leibnizian metaphysics was meant to be the description of a decidedly nonmetaphysical system. "Living animated nature," Buffon argued, "instead of composing a metaphysical degree of beings, is a physical property, common to all matter."[109] The mechanics of reproduction, moreover, were modeled as much on nonorganic "growth" as anything else. Arguing that an individual is "a compound of an infinity of resembling figures and similar parts . . . which can expand in the same mode according to circumstances, and form new bodies, composed like those from when they proceed," Buffon took the cases of crystal growth and vegetation to be paradigmatic:

> We have no other rule to judge by than experience. We perceive that a cube of sea-salt is composed of other cubes, and that an elm consists of other smaller elms, because, by taking an end of a branch, or root, or a piece of the wood separated from the trunk, or a seed, they will alike produce a new tree. It is the same with respect to polyps, and some other kinds of animals, which we can multiply by cutting off, and

separating any of the different parts; and since our rule for judging in both is the same, why should we judge differently of them?[110]

For Buffon, these examples of replication perfectly described growth as a process of mechanical addition and expansion.[111] Thus while organic molecules were deemed living matter—even matter full of "small individuals of the same kind"—it was matter devoid of anything like Leibniz's entelechy.[112]

Sexual reproduction in higher animals required special elaboration so far as the replication now entailed molecules from both parents, and Buffon welcomed the point as an opportunity to rehearse and reject preexistence theories of generation. Like Patrick Blair, Buffon took joint inheritance to be both obvious and testable, and he was at pains when leading up to his discussion of sexual reproduction to rehearse and criticize the leading versions of preexistence theory. The inclusion of polyps—a reference to Abraham Trembley's stunning discussions of the regenerating possibilities afforded freshwater hydra—and "some other kinds of animals" was meant to be a dismissive gesture, since adherents of preexistence theory were initially at a loss to explain the phenomenon.[113] In Buffon's account of sexual reproduction, contact between male and female fluids—the latter referring to a false interpretation of what were in fact Graaffian follicles—began the process of organization leading up to the fully formed fetus. Although Buffon's description of this process is sometimes referred to as one of mechanical (versus vital) epigenesis, there was in fact nothing like an epigenetic or gradual formation of increasingly heterogeneous parts from an original homogeneous mass in Buffon's account. On the contrary, the organic molecules waiting in the sexual "reservoirs" of the parents were already molded in response to their original location; putting together the embryo was thus like putting together a puzzle, since each "piece" was complete and only waiting its proper placement. It is in this sense that Buffon's position is said to be preformationist so far as the parts were preformed by the parents.[114]

Maupertuis, Buffon, and the Problem of Form

The immediate task facing a nonvitalist account such as Buffon's was to explain the principle of order within the complex system of the embryo. As Albrecht von Haller summed up the problem, "Mr. Buffon needs a force which has foresight, which can make a choice, which has

CHAPTER TWO

a goal, which, against all the laws of blind combination, always and unfailingly brings about the same end."[115] The problem was faced by another of the so-called French Newtonians, Pierre Maupertuis. Maupertuis and Buffon had discussed the problems surrounding generation and inheritance on numerous occasions during Buffon's preparation of the first volumes of the *Natural History*, and Maupertuis read through the initial volumes shortly after they appeared.[116] Maupertuis's political and scientific alliances required greater discretion when it came to discussions of generation, however, and his initial publications regarding this were published anonymously.[117] Like Buffon, Maupertuis turned to the models provided by the attractive forces at work in physics, but he referred also to discoveries regarding chemical affinities. The "Tree of Diana"—a tree-shaped formation resulting from an amalgam of silver and mercury—was as significant for Maupertuis's understanding of generation as Buffon's salt crystals had been for him. The Tree of Diana suggested to Maupertuis not only a continuum of natural laws across inorganic and organic matter but a model of formation through chemical affinity. Organic forces, Maupertuis argued in the *Venus Physique*, could be readily understood by analogy with the attraction described in physics and chemistry. In his words,

> Why should not a cohesive force, if it exists in Nature, have a role in the formation of animal bodies? If there are, in each of the seminal seeds, particles predetermined to form the heart, the head, the entrails, the arms and the legs, if these particular particles had a special attraction for those which are to be their immediate neighbors in the animal body, this would lead to the formation of the fetus. Even though the fetus were a thousand times more complex, if the process above were exact, it would still be formed.[118]

But how could affinities for "immediate neighbors" be enough to explain the organization of what were, according to Maupertuis, nonliving particles into a living organism? It was this question that drove René Réaumur's widely known denunciation of the liberal use of forces to explain any number of natural phenomena:

> Everything has its fashions nor is philosophy itself an exception to it: those occult qualities, those sympathies and antipathies which nobody would have dared to name in physicks fifty years ago, have, since that time, showed themselves again with splendor under the name attraction: although we were never taught what this attraction consisted in, very noble uses of it have been made with regard to the mo-

tions of the celestial bodies; great efforts have been made likewise, to make it serve in general to explain all the phenomena in nature.

We are nevertheless as yet very far from seeing anything that resembles any of the organizations which are to concur towards the formation of our great work: how will attractions be able to give to such and such a mass the form and structure of the heart. . . . What law of attraction shall one imagine for the making of that small bone of the ear, whose figure makes it to be called the stirrup?[119]

"We see with the most glaring evidence," Réaumur concluded, "that in order to arrive at the formation of so complicated a piece of work, it is not enough to have multiplied and varied the laws of attraction at pleasure, and that one must besides attribute the most complete stock of knowledge to that attraction."[120] Réaumur's was a critique with impact, for Maupertuis's next publication introduced an account of organic forces that for many recalled the intelligent monads described by Leibniz.[121] Now arguing that the forces of physics and chemistry could never produce a living organism, Maupertuis described organic forces as ones following different laws altogether. "We must have recourse to some principle of intelligence," Maupertuis explained, "to something similar to what we call desire, aversion, and memory."[122] While the organic forces of desire and aversion still functioned similarly to the chemical affinities responsible for the attractive and repulsive forces at work in the formation of the Tree of Diana, an organic force of memory was meant by Maupertuis to solve the problem of embryological formation, since it explained a particle's awareness of its previous location in the parent's body.[123] The forces were originally given to matter by God, after which, as Maupertuis described it, they functioned mechanically in their operations as properties of matter itself. In the same manner that Buffon's organic molecules operated without entelechy and thus ultimately in contrast to Leibniz's metaphysics, Maupertuis's "intelligent" particles were closer to unintelligent replicating machines, such that monstrous births, for example, could now be explained as cases of poor memory on the part of the organized particles. Without undermining Leibniz's importance as a model for Buffon and Maupertuis, therefore, it was clear that both were determined to eliminate any metaphysical role played by a soul in their respective systems.[124]

For Buffon, the task of organization belonged to the combined effect of the interior moulds and the penetrating force, and his discussion recalled the strategy employed in his solution to the problem of determining probabilities in the *franc-carreau* game. Without an entelechy,

CHAPTER TWO

Buffon needed to supply grounds explaining the fitting together of the premolded organic particles, but rather than resort to a principle of intelligence or memory, Buffon concentrated instead on the geometry underlying the position of parts in an embryo. Since "the body of an animal, at the instant of its formation, unquestionably contains all the parts of which it ought to be composed," the main problem was to determine the subsequent expansion (*développer*) of the parts into their final positions.[125] Embryonic expansion, according to Buffon, consisted of two stages. There was a first stage of formation of single parts (head, heart, backbone), which then contained the force to produce doubled parts (arms, legs, ribs) in a second stage of production akin to the processes of vegetation. The mystery lay in discovering the ability of the single parts to determine the specific position of the doubled parts, parts otherwise identical in form: "The left hand is perfectly similar to the right," Buffon explained, "but, if the left hand were placed in the situation of the right, we could not perform the same actions with it."[126] For Buffon this marked the same kind of obscurity faced when looking at a series of symmetrical folds in a paper and trying to determine what the ultimate figure might be:

> We only perceive that the folds [*plicatures*] are uniformly made in a certain order and proportion, and that, whatever is done on one side, is also done on the other. But to determine the figures which may result from the expansion of any given number of folds, is a problem beyond the powers of geometry. The science of mathematics reaches not what immediately depends upon position. Leibnitz's art of *Analysis situs* does not yet exist; though the art of knowing the relations that result from the position of things would be, perhaps, more useful than that which has magnitude only for its object; for we have more occasion to be acquainted with form than matter.[127]

Buffon's comment was certainly prescient given that topology is today used in much this manner when discussing morphogenesis; at the time, however, it left the problem regarding the assemblage of parts unsolved.[128] In 1765 Buffon returned only to the question of generic formation, avoiding any details when describing the protean character of the organic molecules. Species lines remain stable, Buffon now argued, because the molds were proportional to the amount of matter requiring formation. But, he noted, "If this matter were redundant, if it were not at all times equally occupied, and entirely absorbed by the moulds which already exist, it would form others and produce new species. Being alive, it never remains without action; and its union with brute

matter is sufficient to constitute organized bodies."[129] Expanding upon this, Buffon returned to the possibilities of a science based on figure, but now with respect to identifying the source of essential differences between bodies.[130]

Natural History and the History of Nature

Accounting for the generation of individuals was of course only one of the aims of Buffon's *Natural History*. Its largest concern was perhaps the establishment of natural history itself as a classificatory science freed from the province of taxonomy. This effort faced formidable opposition following the spectacular success of Linnaeus's *Systema naturae*—already in its seventh edition—as Buffon began his project.[131] Undeterred, Buffon opened the *Natural History* with a direct attack on the limitations of a system whose groups were determined solely by the parts of fructification. "Who does not see," Buffon demanded, "that whatever proceeds in such a manner cannot be considered a science? It is at the very most only a convention, an arbitrary language, a means of mutual understanding. But no real cognizance of things can result from it."[132] Critical of the "bizarre assemblages" in Linnaeus's taxonomy—"the elm and the carrot, the rose and the strawberry, the oak and the bloodwort"—Buffon suggested that the success of such "ridiculousness" could only be due to the fact this it was "presented with a certain appearance of mysterious order and wrapped up in Greek and botanical erudition."[133] Linnaeus's failure went beyond his attention to an arbitrarily chosen set of organs, however, for according to Buffon he had more importantly failed to grasp the essence of nature's chain of being, a chain whose imperceptible nuances would present no end of anomalies, an infinity of "intermediate species and mixed objects" to confound the systematist.[134] In place of this Buffon offered up "a natural history of all things general and particular," a history whose method—Buffon's synthesis of empirical observation, rational reflection, and probability theory—would provide "the complete description and the exact history of each particular thing," including "not only the history of the individual, but that of the entire species."[135] As a descriptive science whose success was wedded to its attention to nomenclature, taxonomy had been susceptible from the start to Locke's critique, and Buffon understood this. Buffon's attack on Linnaeus, combined with the attention he paid to questions of origin, generation, and genealogy therefore reflected at once Locke's lesson regarding the limitations of classification and Buffon's ambition, nonetheless, to turn natural history into a genuinely explanatory science.

By paying attention to the *entire* history of a species, Buffon shifted the focus of classification from an exclusive concern with divisions between groupings to a science that could include the explanation of a connection between groups. Buffon thus insisted upon a twofold approach to nature—an approach addressing both the history of the individual and the history of the species as a whole—and what this approach revealed was nature's bias toward the latter. For "with regard to individuals," Buffon explained, "she knows not number, and views them only as successive images of the same impression, as fugitive shadows, of which the species is the substance."[136] While the internal molds preserved the special creation of species lines, Buffon took variations to be subsequent creations, representing, in his terms, the "degeneration" of a species. Superficial changes within a species line could be effected as a result of climate or air. Here Buffon's examples relied on changes in climate: a black African moving to Denmark should in the course of time turn white, and the white European moving to Senegal should after the course of long generations develop the skin, hair, and eyes of the African—processes that in either case could be more quickly effected through the "mixture of races."[137] But these were merely changes in color and therefore superficial, according to Buffon, real change regarding form had to be the result of effects on the molds themselves. For this effect Buffon pointed to food, since this was the source of the organic particles with all their protean qualities. And the best example of the degenerating effects resulting from a change in climate and food was the domestication of animals. "Let us compare our pitiful sheep with the mouflon from whom they derived their origin," Buffon began. "The mouflon is a large animal. He is fleet as a stag, armed with horns and thick hooves. . . . How different from our sheep, who subsist with difficulty in flocks, who are unable to defend themselves by their numbers. . . . Timidity, weakness, resignation, and stupidity, are the only melancholy remains of their nature."[138] "The Degeneration of Animals" in 1766 thus took degeneration to be any formal deviation from an original line, a line that was invariably described as stronger, healthier, and more resourceful. Degeneration could in principle be reversed, however, if a species' geographical origin could be determined, the individual relocated there, and sufficient time was allowed to pass.

As Buffon reflected upon the processes of degeneration, he realized that fertile hybrids could serve as clues for reconstructing the genealogical histories of degenerated lines. This manner of investigation, however, lent an air of potential confusion, if not actual inconsistency, to Buffon's account, since he had by then become known for "Buffon's rule," namely, the interfertility criterion as a baseline for the determination of species. Appealing to fertile hybrids suggested to Buffon's readers that he had taken the rule to be breakable after all. What was he up to? The answer lay in the position advanced by Buffon in his essay "Two Views of Nature." In 1765, Buffon was interested in applying his rule to the living history of a species as a whole, and the key to this would be positioning hybrids as a kind of intermediate species. The 1766 essay on degeneration concentrated in part, therefore, on the possibility of fertile hybrids. "The mule," Buffon explained, "which has always been considered as a vitiated production, as a monster composed of two natures, and for that reason has been thought to be incapable of reproduction, is not, however, so deeply injured as has been blindly imagined; for it is not absolutely barren and its sterility depends upon certain external and peculiar circumstances."[139] The reproductive organs of the mule appeared to be as sound as any other animal's, Buffon reasoned, so the cause of its sterility had to lie elsewhere. Considering the many cases of fertile hybrids—Buffon referred to these also as "mules"—produced by different species of birds, Buffon decided in the end that although there might be special difficulties unique to the union of horse and ass, rather than assuming barrenness to be therefore a common feature of hybrids, one should see that "mules [hybrids] in general, which have uniformly been accused of sterility, are neither really nor universally barren."[140] What a fertile mule demonstrated, moreover, was the genealogical connection between the histories of the horse and the donkey. "Under this point of view," according to Buffon,

> the horse, the zebra, and the ass, are all of the same family. If the horse is the principal trunk, the zebra and the ass are collateral branches. The number of their resemblances being infinitely greater than that of their differences, they may be regarded as constituting but one genus, of which the chief characters are apparent, and common to the whole three. . . . Though they form three distinct species, they are not absolutely separated, since the jack-ass produces with the mare, and the horse with the she-ass; and it is probable, that, if the zebra were tamed, he would likewise produce with the mare and the ass.[141]

Buffon went on to rehearse the list of families with "a principle and common trunk from which different branches arise"—sheep and goats, foxes and wolves—distinguishing them, however, from "detached species," that is, species like humans, elephants, and rhinoceroses, which propagated without collateral branches and thus represented both genus and species at once. By the end of his review, Buffon had traced some two hundred species to their origin or point of "ancient degeneration" until fifteen genera or "trunks" and nine detached species had been found.[142]

With Buffon natural history thus became an attempt to grasp a living nature, to grasp species across time and, as a consequence, to base the classification of species upon genealogy. This marked a dramatic transformation in the history of a discipline that until then had been first and foremost a science oriented by its search for the means of discovering nature's divisions and, for that reason, not at all by the patterns of its underlying unity. Buffon's volumes on natural history would quickly come to define what it meant to study nature, and their widespread popularity, their rapid appearance in German and English translations, was due not only to the great accessibility of Buffon's style but to the willingness of a literate public to reconsider their understanding of the basis of organic life altogether. In this manner Buffon's work came to provide not only a method of seeing for scientists but a lens for the imagination when considering nature, and it was as such that it marked the beginning of the revolution that came to place organicism at the heart of both science and the arts in the mid- to late eighteenth century.

THREE

Kant and the Problem of Origin

Kant's Eclecticism

It can come as something of a shock to discover that a good thirty-four years are included under the rubric of Kant's "precritical period," the period covering Kant's work prior to the publication of the *Critique of Pure Reason* in 1781. Although scholars have certainly been right to focus for the main part on Kant's great achievements between 1781 and 1790, it is often forgotten that the bulk of Kant's academic career already lay behind him in the 1780s. From his first publication on the problem of living forces in 1747 to his second attempt to understand the origin of races in 1777 and the appearance of the first *Critique* four years later, Kant had led the life of a busy academic, of a professor with numerous publications and a heavy teaching load. Studies devoted to recovering something of this wide-ranging period have, however, been lately on the rise, and while Kant's various commitments to either Newtonianism or metaphysics have been slightly favored in this literature, the broadest view shows him to have spent much of this period as a dedicated "eclectic."[143]

Open to ideas and social discourse, during the 1750s and 1760s Kant deliberately avoided dogmatic attachment to the views of any one thinker, ultimately preferring to maintain a kind of mitigated skepticism until the proper method for metaphysics could be found. Kant's student Herder characterized this period, for example, as the time of Kant's greatest flourishing, observing that while he "examined Leibniz, Wolff, Baumgarten, Crusius, and Hume, and investigated the laws of nature of Newton, Kepler, and the physicists, he comprehended equally the newest works of Rousseau . . . and the latest discovery in science. He weighed them all, and always came back to the unbiased knowledge of nature and to the moral worth of man."[144] The portrait painted by Herder and other students from Kant's early years as a lecturer show him to have been widely curious, a picture that is indeed consonant with Kant's own self-characterization from that time. "I am myself," Kant wrote, "by inclination a seeker after truth. I feel a consuming thirst for knowledge and a restless passion to advance in it, as well as a satisfaction in every forward step" (20:44).

In light of such wide-ranging eclecticism, it becomes clear that any attempt to precisely capture a sense of Kant's "development" during the early precritical years must fail, so far as the very notion of development presumes a specific end toward which Kant was tending. As Kant's biographer Manfred Kuehn rightly states, "There is no such final goal toward which the early Kant developed. His critical philosophy represents—as he himself tells us—the beginning of something new. It was the result of a sudden, decisive, and radical change in his philosophical outlook, not the fruit of a long, focused search."[145] That part of the story, the one initiated by the "radical change" in Kant's views, has naturally been the source of much speculation. For between Kant's report of "a deep indifference towards my own opinions as well as those of others" (10:74) in 1768 and the appearance of his *Inaugural Dissertation* in 1770, a narrative more closely matching a telic course of development had indeed begun. "The year 1769," Kant recalled, "brought a great light" (18:69), and researchers have sought ever since to discover the source of its illumination, insisting, when necessary, on a prior arc of development to support whatever interpretation is at hand.

At the risk of taking such an approach, I want to address the manner in which Kant was not only open to but actually drawn by questions of origin during the 1750s and 1760s. From the very start of his writing career Kant was interested in a number of theories actively being discussed with respect to cosmological origins. These were also, however, years of lively debates surrounding the problem of understanding biological origin, and Kant would soon develop an interest in these as well, an interest based in part on his appreciation for the intellectuals involved in these debates. Take the case of Maupertuis. When Maupertuis arrived in Berlin to take over the Berlin Academy of Sciences in 1745, for example, he was famous for his Lapland expedition, a trip taken to prove Newton's theory against Descartes's regarding the shape of the earth. But as the suspected author of *Venus Physique* as well, Maupertuis was seen by many to have also begun to stake a place in the controversies regarding biological generation. Kant deeply admired Maupertuis's work on cosmology—both the older discussion of the shapes of the stars (1732) and the new *Essay on Cosmology* (1750)[146]—and he kept careful track of events in the intellectual life of Berlin and the academy. It is therefore no surprise to discover that Kant was also familiar with Maupertuis's work on generation. By the mid-1760s Kant appears in fact to have been well versed regarding the various strategies undertaken to explain biological origins, frequently referring to the central players—Boerhaave, Stahl, Maupertuis, and Buffon—and offering two extended discussions of specific problems facing these accounts in his works from this period.[147]

Despite the fact that Kant kept abreast of developments in the life sciences during the 1750s and 1760s, however, he was pessimistic regarding any possibility of progress in generation theory; discussions of embryogenesis, on Kant's view, simply exceeded the limits of our claims to knowledge of such things. Although Kant's initial judgment on this matter might simply have been reflective of a kind of mitigated skepticism in line with Locke's agnosticism on the matter, by the mid-1760s Kant's stance toward such questions became increasingly tied to a separate problem regarding the origin of knowledge. Between 1765 and 1772 the constellation of issues surrounding this epistemic problem—and questions regarding the origin of ideas in particular—would coalesce into Kant's so-called critical turn, his turn, that is, toward the path leading to the *Critique of Pure Reason*. This turn was inaugurated by Kant's sense that metaphysics must be redefined as a science of limits, of claims limited by the extent and possibilities of our knowledge. But while Kant included the life sciences alongside metaphysics as investigations requiring similar constraint, the eclectic in him was prepared to borrow freely from the models and vocabulary of the embryological debates then underway. Indeed, as I will explain in the following chapters, it was these models that would eventually help Kant discover the origin of knowledge itself.

Matter and Cosmos

Give me matter and I will build a world out of it! (1:230)

Kant's earliest works addressed issues surrounding naturalistic explanations of cosmological origin. Kant's first publication, *On the True Esti-*

CHAPTER THREE

mation of Living Forces (1747), was late to the controversy surrounding Leibniz's account of active force, or *vis viva*, but it was nonetheless a genuine attempt to find some manner of reconciliation between Leibnizians and Newtonians on the question of force.[148] In 1747 Kant was more familiar with Leibniz's metaphysics than Newton's mathematics, and the scale of Kant's reconciliation was clearly tipped toward the metaphysician. By 1754, however, Kant had worked through Newton's system, an account Kant was by then ready to describe as being "as clear as it is indubitable" in his *Spin Cycle* essay of that year (1:186).[149] The *Spin Cycle* essay considered whether the earth's axial rotation had "experienced any change since the earliest times of its origin," and it concluded with a note promising an upcoming essay, extravagantly titled "Cosmogony, or Attempt to Deduce the Origin of the Cosmos, the Constitution of Celestial Bodies, and the Causes of their Motions, from the General Laws of Motion of Matter according to Newton's Theory" (1:191). The promised essay became 1755's *Universal Natural History and Theory of the Heavens*, and Kant was full of confidence regarding the goals he had there set out for himself:

I assert, among all things of nature whose first cause one investigates, the origin of the world system and the formation of the celestial bodies together with the causes of their motions is the one which one may hope to grasp first in a fundamental and satisfactory way. . . . It seems to me that one can here say in a sense without presumption: *Give me matter, I will build a world out of it!*, that is, give me matter, I will show you how a world must arise from it. For if there is matter available which is endowed with an essential attractive force, then it is not difficult to determine those causes which can contribute to the arrangement of the world system, considered at large. (1:229–230)

The general idea behind such a "nebular" hypothesis for the formation of the cosmos—an account according to which attractive and repulsive forces turned an original chaos of particles into increasingly structured bodies—was not entirely novel in 1755, a fact Kant would have been well aware of given his familiarity with the cosmological theories advanced by Maupertuis and Buffon.

When Maupertuis had arrived in Berlin as the newly appointed president of the Berlin Academy of Sciences in 1745, he had faced resistance, even resentment, from a number of sides. He was French, he was a Newtonian in an academy dominated by the metaphysics of Leibniz and Wolff, and he was a clear favorite within King Frederick II's court. Worst of all, Maupertuis seemed determined to remake the academy itself. Reinstituting the academy's annual prize essay question was one thing; demanding that the members actually start working for their pensions was just asking for trouble.[150] One of the first changes instituted by Maupertuis was a broadening of the speculative philosophy class into a class whose discussions could consider material bodies alongside the traditional themes devoted to God and the immortality of the soul. In his initial review of the academy, Maupertuis had felt that this class, above all, was failing, and as an example of the kind of work he hoped to see produced by it, he published his own essay, "The Laws of Motion and Rest Deduced from a Metaphysical Principle," in the first of the academy's publications under his presidency.[151]

Maupertuis had been interested in what he described as "metaphysical mechanics" since his first formulation of what would come to be known as the "principle of least action" in 1744.[152] Explaining that "whenever there is any change in nature, the quantity of action necessary for that change is the smallest possible," what Maupertuis wanted from his essay for the academy was to demonstrate the manner in which a metaphysical principle could be mathematically expressed.[153] In 1744, Maupertuis had applied a mathematical formula to the motion of light in refraction; in the Berlin essay he took on the *vis viva* controversy, arguing that both sides could be comprehended under the same mathematical formula given that the principle of least action could function equally well in describing elastic and nonelastic collisions. What made the 1746 essay appropriate as a species of "metaphysical mechanics" from Maupertuis's point of view—and thus appropriate as a new model for the academy's speculative philosophy class—was the essay's emphasis on the *metaphysics* of the principle of least action over its mathematical exposition, an exposition whose formulas were in fact kept entirely separate from the main body of the text.

In 1750 Maupertuis published the essay again, now under the title *Essay de Cosmologie*.[154] The *Essay* (like the original) began with a critique of the argument from design as a proof for God's existence. Arguments such as these, in Maupertuis's view, were at risk from the start given their susceptibility to counterexamples from the work coming out of the natural sciences. It was much better to look for God in a principle underlying nature's operations as a whole than in the intricate details of particular examples like a honeycomb or the human eye. As Maupertuis expressed this,

The spectacle of the universe seems all the more grand and beautiful and worthy of its Author, when one considers that it is all derived from a small number of laws laid down most wisely. Only thus can we gain a fitting idea of the power and wisdom of the supreme Being, not from some small part of creation for which we know neither the construction, usage nor its relationship to other parts. What satisfaction for the human spirit in contemplating these laws of motion and rest for all bodies in the universe, and in finding within them proof of the existence of Him who governs the universe![155]

Kant echoed these sentiments in his *Universal Natural History*, arguing throughout the preface that appeals to natural beauty or design would always fall short of a proof based on natural laws, and concluding, "There is a God for just this reason, that nature, even in a chaotic state, can develop only in an orderly and rule governed manner" (1:228).[156] When describing the "origin of the cosmos," it was Maupertuis's work on nebulous stars that Kant cited in particular, and the comments there and indeed throughout Kant's early writings demonstrated Maupertuis's influence on Kant as he considered questions of cosmology.[157]

Kant's belief in the eventual discovery of general laws supporting a nebular hypothesis was absent, however, when it came to questions regarding biological formation. In a manner reminiscent of Réaumur's criticism, Kant utterly rejected the possibility that organic processes could be explained by means of the same set of attractive and repulsive forces at work in celestial mechanics. Thus when contrasting discussions of celestial origin with the case presented by organic life, Kant explained that in cosmology all of the questions regarding the coincidence or eccentricity of orbital paths could "be reduced to the simplest mechanical causes. But can we claim such advantages," he asked, "about the most insignificant plant or insect?"

> Are we in a position to say: *Give me matter and I will show you how a caterpillar can be created*? Do we not get stuck at the first step due to ignorance about the true inner nature of the object and the complexity of the diversity contained in it? It should therefore not be thought strange if I dare to say that we will understand the formation of all the heavenly bodies, the cause of their motion, in short, the origin of the whole present constitution of the universe sooner than the creation of a single plant or caterpillar becomes clearly and completely known on mechanical grounds. (1:230)

Celestial mechanics, with all their mathematical complexity, were nonetheless a perfectly knowable basis for understanding cosmological construction. Organic construction, by contrast, could not be grasped through mechanical laws. For, regarding this kind of construction, as Maupertuis had nicely put it, the part-whole relationship was fundamentally obscured by the fact that for any given part "we know neither its construction, usage, nor relationship to the other parts."[158]

Kant's caution regarding organic complexity contrasted with the optimistic tone of another text that Kant found highly useful during this period, the German translation of Stephen Hales's *Vegetable Staticks*.[159] It is clear that Kant was well acquainted with the many experiments described by Hales in his book.[160] Indeed, Kant's master's thesis, *A Succinct Outline of Some Meditations on Fire* (1755), while considered a relatively unambitious treatise, closed in fact with an attempted correction of Hales's consideration of heat. Kant made ready use of Hales's mechanical approach to physiology, for the topics under consideration by Kant at the time—the absorption of water, the properties of air, the cause of earthquakes—were especially suited to such analyses. That said, Kant might have found Hales's book to be oddly satisfying for other reasons as well, for the German translation of Hales's *Vegetable Staticks* unwittingly gathered together much of Kant's own conflicting interests and attitudes toward the investigation of nature at the time.

As a direct translation from the French edition, the German volume included a translation of Buffon's preface. Composed in 1735, Buffon's preface was positively buoyant regarding the possibilities entailed by applications of the Newtonian method. "The furnishings of nature may well rest on distinct parts," Buffon admitted, and "their distinctness remains as unknown to us as their interconnection," but in spite of this, he continued, "how could one measure these secrets with only the imagination or some other such means of discovery, and how could we forget that we find nothing else before us than effects and that these alone should be the means for researching into their causes? Only through precise and correct reflection on constant experience have we ever compelled nature to reveal her secrets. Indeed, one should never strike out on a different path, and all true investigators of nature should regard previous descriptions as nothing other than old dreams."[161] Such positivism contrasted sharply with the attitude taken in the separate preface that had been prepared by Christian Wolff for the German edition. Wolff, for his part, focused his remarks on the problems facing natural scientists, arguing that "knowledge of nature requires a completely different approach, if one is to attain it, than that of those sciences which can be grounded through the correct use of the understanding." Only mathematics, according to Wolff, could be grounded in this manner; as for nature, "Nature lies before us as an abyssal sea and shows us at all places we choose to look the immeasurability of its creator [*Urheber*]." The special difficulty facing natural investigators, as Wolff saw it, stemmed from the fact that "we have only a sensible image of the world emerging from the soul, and it is one which, as a result of the extremely limited powers of the senses, is understood through countless confusions, endlessly repeated anew."[162]

This limitation on the part of the senses put all of the natural sciences at a disadvantage, according to Wolff, and those developed in the wake of Newton's physics were no exception. Wolff rehearsed the failed sciences prior to Newton—Scholastic nominalism, Cartesian substance theory—before turning to the contemporary appeal of attractive and repulsive forces. In Wolff's opinion, Newton was in fact himself to be praised for his adherence to mathematics, it was the Newtonians who deserved criticism. Newton's ideas had simply proven too provocative to resist. "So is it any wonder," Wolff asked, "that the attractive and repulsive forces—which in fact have no basis in matter at all but are rather determined by God himself to be hidden properties whose operations are inconceivable and thus impossible to explain in a clear manner—that these forces should receive such applause in our otherwise enlightened times?"[163] The only recourse for true investigators of nature was to shy away from such "harmful prejudices" and look rather to the resources offered up by metaphysics (*Grund- und Seelenwissenschaft*) when searching for the makeup of material things. What about the Newtonian physiologist Stephen Hales? Hales's experiments had earned him the right to be translated into German, but Hales himself, "in a manner common to his people, more often than not had taken refuge in the attractive and repulsive forces of matter" when interpreting his results.[164] Not to worry, Wolff concluded: reading Hales could not harm the appropriately oriented researcher, for the discussion of forces ultimately demonstrated what they were and remained in the end, mere appearances and phenomena whose origin would never be sensibly discerned.

The juxtaposition of these two prefaces, combined with Hales's own careful descriptions of his work—descriptions that in fact typically avoided any expansive reflections on Hales's part—offered perfect testimony to the uneasy relationship between Wolffian metaphysics-based science and the mechanico-mathematical approach taken by the Newtonians. Whatever inspiration Kant drew from Newton and Maupertuis when discussing cosmological origins in the 1750s, in other words, would have been neatly balanced by Wolff's attitude toward the inscrutability of natural processes. Accommodating both of these in the manner of the *Universal Natural History*, where cosmological origin was reconstructable but biological origin was not, was thus an act performed by Kant entirely in the spirit of the times.

Kant faced another set of competing positions when reading Buffon. More than to either Maupertuis or Hales, it was to Buffon's early volumes of the *Natural History, General and Particular* that Kant turned in the precritical years.[165] The German translation of Buffon's *Natural History* was undertaken by Abraham Kästner between 1750 and 1774, but it was indelibly linked to the famed Swiss physiologist Albrecht von Haller, who had prepared two prefaces of his own for the German edition. Haller's essays were highly critical of Buffon's approach, particularly with respect to his theory of organic generation, and Buffon's failure, as Haller saw it, to account for a principle guaranteeing organization. Thus, after rehearsing Buffon's discussion of internal molds and the penetrating force, Haller complained that these could not provide a reasonable source of organization given the complexity of the body. "In brief," Haller concluded, "what is the cause which arranges the human body in such a way that an eye is never attached to the knee, an ear is never connected to the hand, a toe never wanders to the neck, or a finger is never placed on the extremity of the foot"?[166] Such was the nature of Haller's complaint, but at this juncture Kant's own immediate interests in Buffon were linked to volume 1 of the *Natural History*, the volume Buffon had devoted to questions of cosmological origin and the processes of geological formation.

The clearest sign of this was in Kant's advertisement for a new series of lectures on physical geography for the spring semester of 1757 that were clearly modeled after Buffon's account.[167] In his announcement, Kant promised to clarify and discuss the realms of minerals, plants, and animals—the latter including a promised comparison of the differences in structure and color between men from different regions of the earth—and Kant's proposed list of authors to be assessed for their cosmological theories listed the same men whose views had been discussed at the outset of Buffon's own presentation.[168] The only difference, of course, was that Buffon had now been added as a name to Kant's own list of theorists under consideration. It was in volume 1 that Buffon had discussed the formation of planets, the role played by comets, and the effects of wind and water on geological formation. Buffon's intimation, moreover, of a nebular hypothesis would certainly have been noticed by Kant. "The planets," Buffon had written, "move round the sun in the same direction and nearly in the same plane, the greatest inclination of their planes not exceeding 7 ½ degrees. This similarity in the position and motion of the planets indicates that their impulsive or centrifugal forces must have originated from one common cause," a position that was surely welcomed by a young author of his own *Universal Natural History*.[169]

CHAPTER THREE

By 1757 Kant had already published a treatise on cosmology, a theory of the winds, investigations into the causes of earthquakes, a consideration of whether the earth was aging, and an attempt to discover whether its axial rotation might have changed. His 1757 course announcement accordingly described sections to be devoted to a history of the winds, an account of the seasons both at home and abroad, discussion of rivers and seas, the formation of land masses, and an inquiry into any dramatic changes undergone by the earth in its history. At the end of this, Kant included a short essay considering the relationship between the sea and the moisture in the westerly winds over Königsberg.[170]

It is worth noting that Kant's Latin works during this period were written mainly for the fulfillment of degrees: the *Meditations on Fire* (1755) was Kant's master's thesis, the *New Elucidation of the First Principles of Metaphysical Cognition* (1755) his dissertation, and *The Use in Natural Philosophy of Metaphysics Combined with Geometry. Physical Monadology* (1756) his second dissertation or "*Habilitationsschrift*." The latter pieces especially demonstrated Kant's sensitivity to the range of attitudes toward metaphysics within the sciences. By 1770 Kant would reject monads as much as he would the Newtonian conception of space, and his attitude toward investigations into cosmological origins would be in line with the pessimism expressed all along toward biological theories. But in this earliest part of his career, it was perhaps Kant's eclecticism that most left him open to the possibility of a rapprochement between metaphysics and the physical sciences he so clearly admired.[171]

The Spectacle of Life

It is astonishing that something like an animal body should even be possible! (2:152)

The 1760s identify a period of change so far as Kant's comments regarding biological origin are concerned. While Kant had continued to concentrate on the themes first advertised for the physical geography course in 1757, by the mid-1760s he had clearly also become well versed in the discussions underway between the various schools within the life sciences regarding generation. Kant owned a 1761 German translation of Maupertuis's *Essai sur la formation des corps organisés*,[172] Buffon's volumes discussing generation had been translated into German since the early 1750s, and by the early 1760s Kant would have been well aware of the public back-and-forth between advocates of the preformationist views held by Haller and partisans of the organic force and *vis essentialis* promoted by Caspar Wolff.[173] Although Kant refrained from any work exclusively devoted to the problem, it clearly occupied him enough to receive special treatment in his two longest pieces from this period. Given the context of Kant's remarks—in discussions regarding the decidedly otherworldly topics of God and spirits—Kant's reflections on the problems posed by organic life could almost have been seen as interruptions. What the contexts revealed, however, was the intimate connection, in Kant's view, between attempts to discover a "principle of life" within natural organisms and the search for something beyond the limits of the everyday world. Kant's reflections on generation presumed an audience that was both generally knowledgeable and up to date regarding the latest theories of biological origin, a fact that demonstrates well the kind of fluidity that was then common between the sciences.

In Kant's 1763 essay *The Only Possible Argument in Support of a Demonstration of the Existence of God* he took up again the line of argumentation first advanced in the *Universal Natural History*, even interpolating whole paragraphs from the earlier work.[174] Insisting once more that the regularity of natural causes should be preferred over arguments for God's existence based on design, Kant described what he took be the central weakness of deistic arguments. "Physico-theology," he explained, "regards all the perfection, harmony and beauty of nature as contingent and as an arrangement instituted by wisdom, whereas many of these things issue with necessary unity from the most essential rules of nature" (2:118).[175] Where physicotheology required the contingency of God's choice for its proof—a requirement putting natural laws on the same plane as miracles—Kant argued from the necessity exhibited by natural laws to their a priori ground. It is "the *necessary* unity perceived in *nature*, and the essential order of things, which is in accord with great rules of perfection," for Kant, and only this "leads to a supreme principle, not only of this [God's] existence but indeed of all possibility" (2:116).

On the heels of Kant's comparison of proofs based on contingency versus necessity, Kant turned to the topic of generation as a potentially separate demonstration of the advantages to be had by the naturalistic approach. An argument based on the necessary unity of nature should be enhanced by its ability to derive a variety of effects from a single cause, and as examples of this, Kant listed the effects of gravity, the ether, and the implicit principle of symmetry underlying the structure of snowflakes and flowers. "Nonetheless," Kant argued, "nature is rich in another kind of production. And here, when philosophy reflects on the way in which this kind of product comes into existence, it finds it-

CHAPTER THREE

self constrained to abandon the path we have just described" (2:114). In 1755 Kant had been content to simply brush off the attempt to explain "even so much as a caterpillar" (1:230), but this time he evaluated the options.

Remarking that "it would be absurd to regard the initial generation of a plant or an animal as a mechanical effect incidentally arising from the universal laws of nature," Kant considered in turn the top two competing theories of generation. The first was preexistence theory, according to which each individual being was formed at the time of creation. Historically there had been various interpretations of the specific location of these individuals—though most theorists started with Nicolas Malebranche's encasement or "Russian doll" model—and there were different theories regarding the specific means by which these preformed individuals would transition or "unfold" into normally sized infants. All preexistence theories, however, shared a belief in God's agency in the production of each individual and represented, at the same time, an effort to make room for a mechanical account of the individual's eventual augmentation.[176] Such a view, as Kant understood it, demanded that "each individual member of the plant and animal kingdoms is directly formed by God, and thus of supernatural origin, with only the reproduction [*Fortpflanzung*], that is, only the transition from time to time to the unfolding [*Auswicklung*] of individuals being entrusted to a natural law" (2:114).[177] The second theory Kant considered appealed to God's original agency when producing species lines—a type of generic preformation guaranteeing the reproduction of kinds—but argued for the subsequent generation of individuals according to natural means.[178] Is it possible, Kant asked when introducing this option, that "some individual members of the plant and animal kingdoms, whose origin is indeed directly divine, nonetheless possess the capacity, which we cannot understand, to actually generate [*erzeugen*] their own kind in accordance with a regular law of nature, and not merely to unfold [*auszuwickeln*] them?" (2:114).

Kant went on to rehearse positions that would seem to be examples of this, all the while critical of the specific attempts made in each case to provide a *mechanical* description of the natural means by which individuals would be subsequently generated. Starting with what could easily be described as an allusion to the position held by Hales, Kant also included Buffon and Maupertuis in his critical review:[179]

> It is utterly unintelligible to us that a tree should be able, in virtue of an internal mechanical constitution, to form and process its sap in such a way that there should

arise in the bud or the seed something containing a tree like itself in miniature, or something from which such a tree could develop. The internal forms proposed by *Buffon*, and the elements of organic matter which, in the opinion of *Maupertuis*, join together as their memories dictate and in accordance with the laws of desire and aversion, are either as incomprehensible as the thing itself, or they are entirely arbitrary inventions. (2:115)

But while Kant rejected such accounts as "utterly unintelligible" and "entirely arbitrary inventions," he was equally resistant to the first hypothesis and its recourse to a supernatural origin for every individual member of a species. On this theory human investigation was completely foreclosed, though it could be, as Kant remarked, "supposed that the natural philosophers have been left with something when they are permitted to toy with the problem of the manner of gradual reproduction [*Fortpflanzung*]" (2:115). Here Kant might have named Bonnet as a natural philosopher promoting a revised, even "updated" preexistence theory, so far as Bonnet argued that instead of complete individuals only the rudimentary parts or, for Bonnet, the imprint for the species, were contained in the "germs" of an organism. Such revision did not, however, escape the tincture of the supernatural according to Kant, "for whether the supernatural generation occurs at the moment of creation, or whether it takes place gradually, at different times, the degree of the supernatural is no greater in the second case than it is in the first" (2:115). Returning to the "natural order" offered by Buffon and others, what they had was "not a rule of the fruitfulness of nature, but a futile method of evading the issue" (2:115).

What Kant wanted was something different, a means of avoiding a supernatural solution even if all of the mechanical accounts of individual generation had so far failed. Indeed, as Kant wryly observed, an adequate mechanical explanation of fermenting yeast had yet to be found, but that had hardly led people to suggest supernatural grounds for its existence; the case of plants and animals should be no different. Unless one was willing to rely on divine agency, Kant concluded, "there must be granted to the initial divine organization of plants and animals a capacity, not merely to develop [*Auswickelung*] their kind thereafter in accordance with a natural law, but truly to generate [*erzeugen*] their kind" (2:115). In spite of this, Kant simply could not include organic generation as an example of natural laws at work—and thus, given the larger context of his discussion, as a case of necessity in support for God's existence—for unlike the demonstrable laws guiding cosmological construction, the structure of plants and animals appeared to be unconstrained or contingent while still being oriented somehow toward particular ends.[180]

As Kant neared the end of *The Only Possible Argument*, he returned to the case of organic life, conceding that even this should become understandable in light of God's existence as the ground of all reality and possibility. "And yet," Kant exclaimed, "some amazement is left over. . . . For it is astonishing that something like an animal body should even be possible." Even if a complete mechanical account of the internal "springs and pipes" of the body were available,

> I should still continue to be amazed—amazed at the way so many different functions can be united in a single structure, amazed at the way in which the processes for realizing one purpose can be combined so well with those by means of which some other process is attained, amazed at the way in which the same organization both maintains the machine and remedies the effects of accidental injuries, amazed at the way in which it is possible for a human being to be both so delicately constituted and yet capable of surviving for so long in spite of all the numerous causes which threaten its well-being. (2:152)

It was the unity of purposes within organic life, the fact that organisms could be both self-sustaining and vigilant regarding the need for repair, that made natural products amazing, not the mechanical operations themselves. For Kant it was thus the principle of life, the capacity for a being's generation and self-organization that needed explaining, and recourse to neither supernatural nor purely mechanical grounds of explanation could satisfy that need.

A sense of Kant's continued amazement at the spectacle of organic processes would be provided later by his *Dreams of a Spirit-Seer Elucidated by Dreams of Metaphysics* (1766). After a brief discourse on "A Tangled Metaphysical Knot which can be either Untied or Cut as One Pleases"—a title recalling nothing so much as (and thus presenting a challenge to) the opening passages of Wolff's preface to Hales—Kant took up an argument supporting our "community with spirits," a community based on the possibility that human souls were independent of their bodies.[181] "I must confess," Kant disclosed, "that I am very much inclined to assert the existence of immaterial natures in the world, and to place my own soul in the class of these beings" (2:326). And the grounds for this, he explained, could be equally attributed to the case presented by nature: "The reason which inclines me to this view is very obscure even to myself, and it will probably remain so, as well. It is a reason which applies at the same time to the sentient being of animals.

The principle of *life* is to be found in something in the world which seems to be of an immaterial nature. For all *life* is based upon the inner capacity to determine itself voluntarily [*nach Willkür*]" (2:327). Distinguishing between the "spontaneous activity" of human spirits and whatever inner force might be animating animals, Kant nonetheless took the general point of comparison to be the same in each case: evidence of a principle of life. But how far was this principle to go? Reasoning that even simple elements of matter required some kind of inner activity in order to produce external effects, Kant decided that it would still be "forever impossible to determine with certainty how far and to which members of nature life extends, or what those degrees of life, which border on the very edge of complete lifelessness, may be" (2:330). In spite of this, and thus before turning to the direct discussion of a community of souls, Kant took the time to examine the options when considering the parallel case presented by organic life.

Hylozism could be contrasted with materialism, Kant observed, as tenets arguing that everything is either alive or dead. "Maupertuis," for example, "ascribed the lowest degrees of life to the organic particles of nourishment consumed by animals; other philosophers regard such particles as nothing but dead masses" (2:330). But while it was true that the clear hallmark of life was free activity, Kant argued, it was wrong to say with the mechanists that plants were not thereby alive, for "even though such a being contains within itself a principle of inner life, namely, vegetation, it does not need an organic arrangement to be made for external voluntary activity" (2:331). And even Aristotle's notion of souls—of things divided between the nutritive, perceptive, and rational natures of all beings capable of reproduction and growth—though "probably not capable of proof, was not for that reason absurd" (2:331).

As he had argued in 1763, Kant took the problem facing naturalists to really consist in the impossibility of providing a satisfying account of natural processes, of providing an account, in other words, that relied on neither mechanical nor supernatural explanations. If mechanism or supernatural agency were to be the only options, then, for Kant, even a vitalist like Stahl would be preferable. As Kant put it,

I am convinced that *Stahl*, who is disposed to explain animal processes in organic terms, was frequently closer to the truth than *Hoffman* or *Boerhaave*, to name but a few. These latter, ignoring immaterial forces, adhere to mechanical causes, and in so doing adopt a more philosophical method. This method, while sometimes failing of its mark, is generally successful. It is also this method alone which is of use in science. But as for the influence of incorporeal beings: it can at best be acknowledged to exist; the nature of its operation and the extent of its effects, however, will never be explained. (2:331)

The "influence of incorporeal beings," like the vegetative and animal souls described by Aristotle, were simply incapable of proof, and, in the absence of their explicit demonstration, they should thus simply "be acknowledged to exist." To be clear, Kant was not interested in undermining the value of mechanical description here; on the contrary, it was *the* method of science, without which science could lapse into a dogmatic recourse to occult forces and miracles when explaining natural phenomena. It was just that, for Kant, the life-matter relationship was perfectly parallel to the problem of understanding the mind-body relationship: in each case we were ignorant not only of the manner in which they were united, but of the very nature of life and souls themselves. It was presumption on the part of metaphysicians and natural scientists alike to forget this ignorance, a forgetting that could lead to all manner of "metaphysical knots."

As Kant worked on *Dreams of a Spirit-Seer*, he continued to teach a heavy schedule of courses. Since 1757 Kant had been offering lectures on physical geography with enough success, and clearly enough interest on his own part, to announce in his course description for 1765–1766 that he would henceforth be dividing this course between a condensed discussion of "physical, moral, and political geography" and an expanded consideration of man, an investigation promising "a comprehensive map of the human species" (2:313). This second part would eventually become an independent course on "anthropology" which, like "physical geography," Kant would continue to teach every year for the remainder of his career.[182] Buffon had closed the third volume of his *Natural History* with an extensive discussion of "The Varieties of the Human Species," and it seems not unlikely that Kant was influenced by Buffon's account when pulling together his own materials for the newly expanded portion of the course.[183]

The care with which Kant was reading Buffon during this period showed itself most clearly, however, in 1768's essay *Concerning the Ultimate Ground of the Different Regions in Space*. There Kant worked in close dialogue with Buffon's appeal to the possibilities of a geometry of position, following him in identifying the starting point for his own reflections as Leibniz's *"analysis situs."* "It looks as if a certain mathematical discipline," Kant began, "which Leibniz called *analysis situs*, and the loss of which was lamented by Buffon among others when he was considering the foldings together [*Zusammenfaltungen*] of nature in the seeds—it looks as if this discipline was never more than a thought in Leibniz's mind" (2:377). In 1768 Kant was right to judge the rumored geometry to have been "never more than a thought," since at that point few people, if any, would have read Leibniz's short discussion of the proposed geometry.[184] Kant took as his starting point, therefore, Buffon's own discussion, with which Kant was obviously familiar. Kant set his own agenda as follows: "What we are attempting to demonstrate, then, is the following claim. The ground of the complete determination of a corporeal form does not depend simply on the relation and position of its parts to each other; it also depends on the reference of that physical form to universal absolute space, as it is conceived by the geometers" (2:381). In his discussion Kant repeatedly rejected a formulation, such as Buffon's, of grounds on the basis of "the manner in which the parts of the body are combined with each other" or the "positions of the parts of matter relative to each other" (2:382, 383), but he nonetheless took his cue for the whole from Buffon's comment regarding the need to explain the body's symmetry of difference.

Buffon had ascribed the process of such differentiation to the mysterious means by which the interior molds seemed to recognize, for example, the difference between left- and right-handedness, a difference significant for the functioning of the organism as a whole. In Buffon's words,

There are many more double than simple parts in the body of an animal, which seem to be produced on each side of the simple parts by a kind of vegetation; for these double parts are similar in form, and different in position. The left hand exactly resembles the right, because it is composed of the same number of parts; nevertheless, if it was placed in the situation of the right, we could not make use of it for the same purposes, and should have reason to regard it as a very different member. It is the same with respect to the other double parts; they are similar as to form, and different as to the position which is connected to the body of the animal; and by supposing a line to divide the body into two equal parts, the position of all the similar parts would refer to this line as a center.[185]

Kant recast this problem as part of an argument for absolute space, since the internal distribution or positioning of parts could in fact only be determined as "left" or "right" on the basis of something external to the body altogether. It was the body's position relative to the surrounding regions of an absolute space—"above and below, right and left, in front and behind"—that allowed us to orient ourselves and even explained the phenomena of "incongruent counterparts" like left and right hands.

Like Buffon, Kant was interested in the promises afforded by topology, but to use the internal relation of parts as a starting point, as Buffon had done, was to eliminate the possibility of a constant orientation. Kant's three-dimensional topology moved the frame of reference outside of the body, which made it "unsurprising that the ultimate ground, on the basis of which we form our concept of regions in space, derives from the relation of these intersecting planes to our bodies" (2:379).

Whereas Buffon had pointed to the mysterious workings of the interior moulds in determining the position of parts, it was nature, according to Kant, that implanted in us a feeling (*Gefühl*) for the difference between left and right. "Since the distinct feeling of the right and the left side is of such great necessity for judging directions," Kant argued, "nature has established an immediate connection between this feeling and the mechanical organization of the body" (2:380). And nature instantiated this difference both physiologically in humans—Kant cited empirical studies by Borelli and Bonnet—and morphologically in species like snails and hops when determining the specific orientation of their curves (2:380).

Buffon may have avoided anything like Maupertuis's recourse to the intelligent monads in his later accounts, but Buffon's reluctance to speculate further had thereby left the problem of guidance unresolved within his theory. For his own part, Kant accepted that "the action of the creative cause in producing the one [hand] would have of necessity to be different from the action of the creative cause producing the counterpart" (2:383), but unlike Buffon, Kant insisted that the "inner ground" of this difference could not "depend on the difference of the manner in which the parts of the body are combined with each other" (2:382). Instead it was "the action of the creative cause" in connecting the mechanical organization of the body to the feeling of right and left—a feeling whose points of orientation were themselves fitted to the surrounding regions of absolute space—that constituted the "inner difference." "Our considerations," Kant concluded, "make it clear that differences, and true differences at that, can be found in the constitution [*Beschaffenheit*] of bodies; these differences relate exclusively to *absolute* and *original space*," and the proof of this lay in the phenomenon of left- and right-handedness or the fact that "the form of a body exclusively involves reference to pure space, and that is by holding one body against other bodies" (2:383).

At the end of the essay Kant used this argument to challenge "German philosophers, according to which space simply consists in the external relation of the parts of matter which exist alongside each other," for, as Kant saw it, Leibniz's position left the fact of incongruent counterparts inexplicable (2:383).[186] But instead of ending with a note of triumph against the "German philosophers," Kant was cautious in closing, acknowledging the difficulty of advancing an argument that attempted to use "ideas of reason" to describe something whose reality was known intuitively by inner sense—a note of caution whose tone, while recent, was not altogether new for Kant.

Two years earlier, in *Dreams of a Spirit-Seer*, Kant had followed his argument in *support* of a community of souls with a chapter called "Anti-Cabbala—A fragment of ordinary philosophy, the purpose of which is to *cancel* community with the spirit-world" (italics mine). As for the principle of life within nature, Kant announced that this was a topic about which nothing could be positively said (2:351). Here Kant was explicit regarding the specific problem at hand: "The various appearances of *life* in nature [*Erscheinungen des Lebens*] and the laws governing them, constitute the whole of that which it is granted us to know. But the principle of this life, in other words, the spirit-nature which we do not know but only suppose, can never be positively thought, for, in the entire range of our sensations, there are no *data* for such positive thought" (2:351). Here was Kant's affirmation of laws governing the appearances of nature, laws capable of being understood according to the physical principles set forth in mechanics, for example. The difference between a principle of life thought to be guiding nature and laws regulating the appearance of nature, however, was the utter lack of any evidence supporting our belief in a principle of life. The experiences of everyday life might yield a sense of nature's vital principles, but what evidence beyond this could secure such a claim?

Intuitions regarding a principle of life faced, in fact, far graver difficulties than "ideas of reason" concerning absolute space. We experience ourselves in space, Kant reasoned, and we intuit its reality through the feeling of our body's orientation within it. In 1768 Kant thought that this subjective emphasis on experience and feeling required a separate demonstration, one based on ideas of reason via the examination of incongruent counterparts. In 1766 Kant had offered no such parallel account of a feeling of life, and however he might have cataloged his experience of astonishment at life's productions, this astonishment remained incomprehensible. Kant might have privately sympathized with Stahl's vitalism over the mechanism of Boerhaave or Hoffman, therefore, but so far as he now understood them, the sciences of life, no less than the sciences of the soul, would have to be reined in if they were to succeed at all.

FOUR

The Rebirth of Metaphysics

A Philosophy Is Born

Before true philosophy can come to life, the old one must destroy itself. (10:57)

When it came to reforming the sciences, Kant's plans were specific. For knowledge to move forward, Kant argued, certain limits had to be set, a circumscription that began with two questions. First, one must ask what kind of knowledge would be required in order to solve a given problem, and second, one must decide whether that knowledge was in fact possible (10:56). When one searched for vital principles or inquired into the character of the spirit world, the objects of investigation were simply unknowable. Kant was, moreover, ready to diagnose and name the exact source of so much error in these sciences. "Surreptitious concepts"—or "subreptive axioms" as they would later be called—described a specific transgression: the crossing of fields of knowledge meant to be separate.[187] Subreption created confusion in the sciences when investigators took concepts gleaned from experience, the experience of magnetic forces, for example, and used them to describe processes that were incapable of experience, processes like embryological formation or the means by which spirits might communicate. As Kant introduced the problem in *Dreams of a Spirit-Seer*, "There are many concepts which are the product of covert and obscure inferences made in the course of experience; these concepts then proceed to propagate themselves by attaching themselves to other concepts, without there being any awareness of the experience itself on which they were originally based or of the inference which formed the concept on the basis of that experience. Such concepts may be called *surreptitious concepts [erschlichene Begriffe]*" (2:321). Kant actually charged himself with having made this mistake as well, given that both his *New Elucidation* (1755) and the *Physical Monadology* (1756) had ascribed the forces of attraction and repulsion to spirits and monads (1:415, 484). As an act of contrition, he was now ready to declare, "It is impossible for reason ever to understand how something can be a cause, or have a force; such relations can only be derived from experience." Indeed, "All judgments, such as those concerning the way in which my soul moves my body, or the way in which it is now or may in the future be related to other beings like itself, can never be anything more than fictions" (2:370, 371).[188]

These were all of course topics that had been the special province of metaphysics, so one immediate problem was to consider what might be left for the metaphysician to do without them. Kant was prepared for this. "Metaphysics," he began, "with which, as fate would have it, I have fallen in love but from which I can boast of only a few favours, offers two kinds of advantage." The first advantage was the ability of metaphysics to aid reason in spying out the hidden properties of things; indeed, when left unfettered, reason was unrivaled in its capacity for such inferences. It was in fact this talent for inferences that had led to the very problem facing metaphysics now. But this was balanced by what Kant took to be the second advantage afforded by metaphysics. As Kant described it,

> The second advantage of metaphysics is more consonant with the nature of the human understanding. It consists both in knowing whether the task has been determined by reference to what one can know, and in knowing what relation the question has to the empirical concepts, upon which all our judgements must at all times be based. To that extent metaphysics is a science of the *limits of human reason*. A small country always has a long frontier, it is hence, in general, more important for it to be thoroughly acquainted with its possessions, and to secure its power over them, than blindly to launch on campaigns of conquest. (2:367–368)

The key here was to see that by redefining metaphysics as a science of the limits of human reason, Kant had radically redirected investigators away from the topics of life and soul toward an examination of reason itself. Once all investigations were to be prefaced by a separate inquiry regarding the abilities of reason, moreover, the task of determining the nature and limits of cognition became both necessary and ultimately identical to the task of determining the nature and limits of scientific investigation itself.

As Kant mapped the contours of this investigation, he took its outcome, at the very least, to be the elimination of surreptitious concepts. "All of my endeavors," Kant explained, "are directed mainly at the proper method of metaphysics (and thereby also the proper method for the whole of philosophy)" (10:56). The method would determine the scope of reason's possibilities—and thereby also the scope of any rational investigation—and eradicate surreptitious concepts as a result. If it was certain that a genuine insight into organic processes was impossible, then the surreptitious appeal to an "irritable force," as made by Haller for example, should be avoided when explaining muscle contraction. This did not mean an end to further investigations in the life sciences, but it did mean, as noted earlier, that only "the appearances of *life* in nature, and the laws governing them, [would] constitute the whole of that which it is granted us to know" (2:351). Naturalists could focus on the regularity of nature's appearances and continue their search for mechanical causation, but the search for vital principles and the attempt to understand the mysteries of generation and reproduction were invariably riddled with surreptitious concepts and thus doomed from the start.

But while Kant was confident in his diagnosis of the need for reform in the sciences, he was still unsure of the task left before him. "My problem is this," he wrote. "I noticed in my work that, though I had plenty of examples of erroneous judgments to illustrate my theses concerning mistaken procedures, I did not have examples to show *in concreto* what the proper procedure should be" (10:56). So far as the "proper procedure" amounted to precisely delineating the limits of reason, what Kant was missing, in other words, was a positive description of reason itself.

Up until now Kant had written extensively upon questions connected to cosmological origin, and, his criticisms notwithstanding, he was thoroughly versed in the leading theories of biological origins as well. As he now took on the job of re-creating metaphysics as a science of the limits of human reason, the first task concerned questions regarding the origins of knowledge. Was it the case, as rationalists had it, that true ideas were like seeds implanted in the soul by God—a strategy in some sense parallel to that adopted by the preexistence theorists—or were empiricists correct instead when identifying the senses as the true origin of ideas? Kant was long familiar with the rationalists' reliance on the intellectual intuition of innate ideas, and as for the other option, the mid-1760s were perhaps the heyday of Kant's engagement with British empiricism.

Kant's closest friend during this period was the British merchant Joseph Green, a merchant known for his literacy in the writings of Hume and other members of the Scottish enlightenment.[189] This was surely a topic of shared interest, for in Kant's description of his ethics lectures planned for 1765–1766, for example, he wrote, "The attempts of Shaftsbury, Hutcheson, and Hume, although incomplete and defective, have nonetheless penetrated furthest in the fundamental principles of all morality" (2:311). We know, moreover, that by the mid-1760s Kant was well acquainted with a 1755 translation of Hume's *Enquiry Concerning Human Understanding* as well. "Since the *Essays* of Locke and Leibniz, or rather since the origin of metaphysics so far as we know its history," Kant later declared, "nothing has ever happened which could have been more decisive to its fate than the attack made upon it by David Hume. He threw no light on this kind of knowledge; but he certainly struck a spark from which light might have been obtained, had it caught some inflammable substance and had its smouldering fire been carefully nursed and developed" (4:257). Indeed, taken as whole, 1766's *Dreams of a Spirit-Seer* reverberated with the potency of Hume's skepticism regarding the "dreams of metaphysics," and there was thus real justice in Kant's eventually citing Hume for having woken him from a dogmatic slumber (4:260). But while Kant's adoption of a skeptical methodology put him in position to recognize a need for reform in the sciences, the work facing him now required him to take a positive stance regarding the workings of the mind, and for this Kant turned first to the work of Leibniz and Locke.

Like Hume's, Locke's works were both available and well known in Germany at this time. Georg Kypke had already been a longtime friend of Kant's when he translated Locke's posthumously published addendum to the *Essay Concerning Human Understanding* in 1755.[190] And by then too Kant would have had access to the new Latin translation of the *Essay* itself, a book he described as "the ground of all true *logica*" (24:37).[191] Indeed, as Kant later explained in his course on logic, "Some books are of great importance and require considerable inquiry; these one must read often, e.g., Hume, Rousseau, Locke, who can be regarded as the grammar of the understanding, and Montesquieu, concerning the spirit of the laws" (24:300). By the end of his career, references to

Locke would be peppered throughout Kant's notes, lectures, and published writings.[192]

In Kant's notes, in particular, Locke would frequently be paired with Leibniz. The two were typically cited for their investigations into the origin of ideas, and in the years after *Dreams of a Spirit-Seer* this made them significant interlocutors for Kant in his attempted reform of the sciences. Although it is unknown when exactly Kant read through the posthumous publication of Leibniz's *New Essays* (1765), it seems likely that by 1770 he was at least familiar with the first page of it, since Leibniz's opening formulation of the divisions of philosophy would be subsequently repeated by Kant on numerous occasions.[193] As Leibniz positioned himself there against Locke,

> Although the author of the *Essay* says hundreds of fine things which I applaud, our systems are very different. His is closer to Aristotle and mine to Plato. . . . Our disagreements concern points of some importance. There is the question whether the soul in itself is completely blank like a writing tablet on which nothing has as yet been written—a *tabula rasa*—as Aristotle and the author of the *Essay* maintain, and whether everything which is inscribed there comes solely from the senses and experience; or whether the soul inherently contains the sources of various notions and doctrines which external objects merely rouse up on suitable occasions, as I believe and as [does] Plato.[194]

Kant accepted Leibniz's divisions, often visually grouping them—Aristotle in a column with the empiricists, Plato with the intuitionists—when writing out his lectures. But where did that leave Kant? In 1769 Kant appeared to be torn as to how to proceed with his investigation into the origin of knowledge, accepting, on the one hand, Locke's dictum regarding sense as the necessary occasion for all thought, and Leibniz's admonishment, on the other hand, regarding the impossibility that a concept of God could ever arise from the senses. As Kant outlined his own view of cognition, the picture thus presented an amalgamation of Leibniz and Locke:

> Some concepts are abstracted from sensations, others merely from the law of the understanding for comparing, combining, or separating abstracted concepts. The origin of the latter is in the understanding; of the former, in the senses. All concepts of the latter sort are called pure concepts of the understanding, *conceptus intellectus puri*. We can of course set these activities of the understanding in motion only when occasioned to do so by sensible impressions and can become aware of certain concepts of the general relations of abstracted ideas in accordance with the laws of

the understanding; and thus Locke's rule that no idea becomes clear in us without sensible impression is valid here as well; the *notiones rationales*, however, arise no doubt by means of sensations and can also only be thought in application to the ideas abstracted from them, but they do not lie in them and are not abstracted from them (*sind nicht von ihnen abstrahirt*). (17:352)

Kant thus remained in keeping with Locke insofar as even the "pure concepts of the understanding" were concepts empirically gleaned through reflection on the contents of sense. These concepts were to be distinguished—as "abstracted" from sense—from those rational notions whose origin would have to be different.

When Kant was offered a chair in logic and metaphysics the following year, he was required to present an "inaugural dissertation," a piece that would offer him the opportunity to fulfill his promised reform of metaphysics. When it was completed, Kant's solution to the problems of origin and subreption stood at the forefront of the project. It was a solution, at least with respect to subreption, that could be accomplished in one stroke by the radical division Kant now proposed between the faculties of sense and intellect. Insisting that the method of metaphysics concern itself wholly with the prevention of "subreptive axioms," Kant argued that only a radical separation between sense and intellect could avoid the possibility of such cross-contamination; a separation effectively announced in the book's title: *On the Form and Principles of the Sensible and the Intelligible World*.

When Kant first broached the problem of subreption in 1764–1765, he had sent his thoughts—including those already mentioned regarding the problem of having nothing more than negative examples to illustrate his point—to J. H. Lambert, a philosopher and mathematician living in Berlin. Lambert had agreed with Kant's call for reform, observing in reply that "whenever a science needs methodical reconstruction and cleansing, it is always metaphysics" (10:62). Turning to the subject of Kant's proposal, Lambert had made a number of remarks:

The first concerns the question whether or to what extent knowing the *form* of our knowledge leads to knowing its *matter*. This question is important for several reasons. First, our knowledge of the form, as in logic, is as incontestable and right as is geometry. Second, only that part of metaphysics that deals with form has remained undisputed, whereas strife and hypotheses have arisen when material knowledge is at issue. Third, the basis of material knowledge has not, in fact, been adequately shown. . . . Fourth, even if formal knowledge does not absolutely determine material knowledge, it nevertheless determines the ordering of the latter, and to that extent we ought to be able to infer from formal knowledge what would and what would not serve as a possible starting point. (10:64)

Lambert's emphasis on the connection between form and matter proved highly influential for Kant, as was clear from the start of Kant's finished text. The *Dissertation* opened, for example, with a discussion of the concept of a world, in particular of "its two-fold origin in the mind" (2:387). While Locke had been correct with respect to the matter of sensations, Kant now argued, form was a result of mental determination and thus lay in the mind. This was not to end up on the side of Leibniz against Locke, however; on the contrary, Kant was suggesting something new. The imposition of form and the supplying of content described a division of labor that would now be applicable to both sense and intellect alike.

Kant started with the case of sensitive knowledge, explaining that in "representations of sense there is in the first place something that we may call *matter*, i.e., sensation [*sensatio*], and something else that we may call *form*, i.e., the sight [*species*] of sensible things, which obtains when various things which affect the senses are co-ordinated by a certain natural law of the mind" (2:392–393). The process by which form was imposed upon sensation followed Locke's model of reflective comparison, a process described by Kant as the result of the "logical use" of the understanding, whereby sensations could be classified or subordinated under common class concepts. Unsorted sensations remained at the level of appearance (*apparentia*), according to Kant, whereas "the reflective cognition which arises from the intellectual comparison of a number of appearances is called *experience*" (2:394). This somewhat borrowed account of sensible cognition sat alongside Kant's genuine innovation, the identification of space and time as the "schemata and conditions of all human knowledge that is sensitive" (2:398).

Two years earlier Kant had argued that space was absolute; now space and time were jointly identified as the formal, yet subjective, principles of the phenomenal world. These were the principles underlying mechanics and geometry, fields that each yielded undisputable truths. With this move Kant repositioned the status of sensitive knowledge. For although he insisted that we remember the sensible origin of even the most abstract laws of sensible phenomena, his insistence concerned the specter of subreption, not the quality of sensible knowledge. "There is thus a science of sensible things," as he put it, one that "yields us quite genuine knowledge, and at the same time furnishes a model of the highest certainty for knowledge in other fields" (2:398). Against a critique like Wolff's regarding the confused perceptions of the senses, therefore, Kant presented a theory of sensible cognition that explained the success of geometry and mechanics—at the same time that it could be validated by that success—and grounded the possibility of certainty regarding sensible phenomena.

Kant's discussion of sensible cognition proceeded in stages. The first concerned Kant's shifting the focus away from objects of perception toward our mental representations of them, since representations alone were susceptible to the mind's imposition of form. A representation, Kant explained, "indicates a certain aspect or relation of the sensa and yet is not properly an outline or schema of the object, but only a certain law inborn in the mind coordinating with one another the sensa arising from the presence of the object" (2:393). The task for Kant was to balance the quality of the real, one granted by the material content of a sensation, with the opportunity for control of that content through the mind's inborn laws. Once laws for the mental construction of phenomena became too thorough, Kant realized, the account would risk charges of idealism (2:397). Leaving these difficulties aside, Kant concentrated instead on his argument for sense certainty.

A genuine knowledge of sensible phenomena was possible, according to Kant, because judgments about sensible objects fell under the purview of the logical use of the understanding, and the logical use of the understanding was concerned only with determining the internal agreement between subjects and predicates in judgments. By focusing, therefore, on the internal relationship between subject and predicate over the supposed, but unknowable, external connection between subject and object, certainty regarding phenomena could be guaranteed by the proper functioning of the mind's laws for construction. "Consider judgments about things sensitively known," Kant began. "The truth of a judgment consists in the agreement of its predicate with the given subject. But the concept of the subject, so far as it is a phenomenon, can be given only by its relation to the sensitive faculty of knowledge, and it is also by the same faculty that the sensitively observable predicates are given. Hence it is clear that the representations of subject and predicate arise according to common laws, and so allow of a perfectly true knowledge" (2:397). Phenomena, for Kant, were thus the synthetic result of sensible matter and the mind's imposition of form, a synthesis accomplished through laws grounding the certainty of sensible experience.

But what, precisely, was Kant's understanding of these laws for the logical coordination of sensible data, laws that were said to be inborn (*innatas*) in the mind? Were they meant to balance Kant's deference to Locke regarding the independent reality of material sensations? For now, at least, Kant left the status of these laws unexamined. Not so the concepts of space and time. These would fall into a third category, somewhere between the sensible acquisition of empirical concepts and the mental recovery of innate ideas; space and time, according to Kant, were "originally acquired" so far as they were generated by the mind itself. Asking rhetorically whether space and time were connate (*connatus*) or acquired (*acquisitus*), Kant immediately rejected the possibility of their empirical acquisition so far as that would render geometry contingent, something it clearly was not. The alternative was "not to be rashly admitted," either, however, "since in appealing to a first cause it opens the path to that lazy philosophy which declares all further research to be in vain" (2:406). Instead, Kant argued that the origin of space and time lay between these alternatives:

Both concepts are without doubt acquired, being abstracted not from the sensing of objects (for sensation gives the matter, not the form, of human cognition) but from the action of the mind in coordinating its sensa according to unchanging laws—each being, as it were, an immutable type to be known intuitively. Though sensations excite this act of the mind, they do not influence the intuition [*non influunt intuitum*]. Nothing is here connate save the law of the mind, according to which it combines in a fixed manner the *sensa* produced in it by the presence of the object. (2:406)

It was clear that much rested, therefore, on the laws of the mind: the regularity of their logical operations generated space and time as the pure forms of sensible intuition, and through their subsequent coordination of sense data, they grounded the possibility of sense certainty.

The notion that concepts might be generated or "originally acquired" through the workings of cognition marked Kant's major advance from the position he had outlined in 1769. At that point, even the "pure concepts of the understanding" fell under the Lockean model of concepts gleaned by abstraction from sense (17:352). On Locke's view, it was in fact this "gleaning," so to speak, that constituted the main work of the understanding. Things had clearly changed for Kant by the time he composed the *Dissertation*. The difference between intellectual concepts—"possibility, existence, necessity, substance, cause, etc. with their opposites or correlates"—destined for the "real use" of the intellect (2:395) and the sensitive concepts of space and time controlled by its "logical use" (2:398) were enormous, according to Kant, but their birthplace was the same. Like the concepts of space and time, Kant considered intellectual concepts to have been given "in the very nature of the pure intellect, not as concepts connate to it, but as concepts abstracted (by attention to its actions on the occasion of experience) from laws inborn in the mind, and to this extent, as acquired concepts" (2:395).[195] The difference between sensible and intellectual concepts lay, therefore, in their objects, not their origin, even if "each kind of knowledge preserves the mark of its descent, so that the former kind, however distinct, is on account of its origin called sensitive, while the latter, however confused, remains intellectual" (2:395).

While the sensitive concepts of space and time grounded an experience of phenomena that was capable of staving off skepticism—"The laws of sensibility will be laws of nature, insofar as nature falls within the scope of the senses" (2:404)—the case was different for intellectual concepts. Sensitive concepts were applied to sensible intuition, the *"apparentia"* waiting to be organized into a coherent experience; intellectual concepts, by contrast, had no intellectual intuition with which to work. This explained the clarity of geometry when compared to the obscurity surrounding the traditional content of metaphysics. Kant emphasized this restriction, moreover, for it was precisely such overreaching that had opened the door to the surreptitious application of sensitive concepts in the first place. "No intuition of things intellectual but only a symbolic knowledge of them is given to man," he declared, for "thinking is only possible for us by means of universal concepts in the abstract, not by means of a singular concept in the concrete" (2:396). While sensation provided direct contact with its contents, the intellect had to work discursively, either through intellectual concepts or through its generation of moral exemplars to guide actions, the exemplars of God and moral perfection, for example. And neither of these possibilities contained the kind of content boasted of by sense. "All the matter of our knowledge is given by the senses alone," Kant concluded, "whereas a noumenon, as such, is not to be conceived through representations derived from sensations. Consequently, a concept of the intelligible as such is devoid of all that is given by human intuition" (2:396).[196]

For someone newly interested in reorienting metaphysics toward an account of the extent and limits of human reason, Kant had thus made a good start. Cognition was now described as a twofold exercise, one that was both sensitive and intellectual. The intellect was described as having both a logical and a "real" use, with the former devoted to the task of logical subordination according to laws inborn in the mind.

CHAPTER FOUR

This subordination was responsible for both the discrimination of sense data in the generation of empirical judgments and the logical exercises associated with reflection on the concepts and exemplars generated by the intellect's real use. "The logical use, but not the real use," Kant explained, "is common to all the sciences" (2:393). By the real use of the intellect, "the very concepts of objects or relations" were acquired through the nature of the intellect itself (2:393). The intellectual concepts generated in this manner provided concepts of objects in terms of their existence, substance, possibility, necessity, cause, and number; in their so-called dogmatic use, they issued moral exemplars.[197] By denying the intellect any content for its intellectual concepts, Kant could argue that he had staved off the path leading investigators to the use of surreptitious concepts. And the attention Kant paid to the difference between an abstractive process yielding empirical concepts and one that could, by its own workings, actually generate or "originally acquire" concepts, identified Kant's new solution to the problem of understanding the origins of knowledge.

From Original Acquisition to the Epigenesis of Knowledge

Given the focus of the *Dissertation*, Kant must have been tempted when that year's topic for the *Preisschrift* was announced: an essay that could reconcile Descartes and Locke on the origin of ideas.[198] But whether Kant's research agenda for 1770 left him inclined to take up the topic or not, the real question is how he had arrived at his solution to the problem. What models were there for his description of an original acquisition of sensible and intellectual concepts? Kant had worked closely with Buffon's text when preparing his account of space, but 1768 was also the year that Kant had reported "a deep indifference towards my own opinions as well as those of others" (10:74). And what resources existed for Kant's discussion of the laws whose workings generated concepts in the first place? Laws like these, or certainly processes with similar functions, were assumed by Locke and Leibniz both; indeed the Leibnizian quip that "for Locke nothing is in the understanding—except the understanding itself" turned on that fact.

As for Leibniz's account, Kant seemed determined to avoid innate ideas, reproaching, on the one hand, such "rash" recourse to innatism as a type of lazy philosophizing and eliminating, on the other hand, the kind of intellectual intuition that would be required for them. Leibniz's innatism had turned in part on connections Leibniz saw between Plato's doctrine of recollection and Leibniz's own sense that "the seeds of the things we learn are within us—the ideas and the eternal truths which arise from them."[199] Like the divinely implanted seeds, the mind too came from God, an origin explaining its capacity to realize eternal truths in the first place. "So it is not a bare faculty," as Leibniz characterized the mind in his *New Essays*, "consisting in a mere possibility of understanding those truths: it is rather a disposition, an aptitude, a preformation, which determines our soul and brings it about that they are derivable from it."[200] As was seen earlier, the preexistence theory of encasement operated for Leibniz as a biological analog to his own theory regarding the formation of ideas. In each case there was a "virgin birth," so far as both individuals and ideas regarding eternal truths were generated from seeds implanted by God at the creation of the world. But if by 1770 Kant wanted something more than Locke's account of empirical concepts abstracted from the senses, it is also clear that he wanted something less than the harvesting of truths grown up from seeds that God had sown in the mind.

While it has been fair game to speculate on the source of the "great light" that the year 1769 brought to Kant (18:69), attending to the problem of origin at least points one past the usual suspects. Leibniz was hardly uncommon in his liberal use of vocabularies drawn from both religious and scientific discourses, and his appreciation for Plato aside, Leibniz's strategy was in fact deeply suggestive of Kant's own solution to the problem of origin. But whereas Leibniz had appealed to preexistence theory as a biological analog, it seems likely that Kant had some form of epigenesis in mind when describing the mind's generation or "original acquisition" of concepts. When Kant proposed in 1763 that we forgo supernatural accounts of generation, and mechanical views as well, he had argued that what science needed instead was an explanation that "granted to the initial divine organization of plants and animals a capacity, not merely to develop their kind thereafter in accordance with a natural law, but truly to generate their kind" (2:115). By 1770, Kant was convinced that such an explanation could come only at the cost of subreption. He seems to have felt, however, that the two-step model of divine formation and organic generation *could* be safely mapped onto a theory of cognition aimed at explaining the generation of concepts from innate laws. The details were still fuzzy. It was not yet clear to Kant, for example, how these concepts were specifically connected to the implanted laws for logical subordination from which they arose, but the strategy epigenesis offered for discovering an origin that was neither supernatural nor empirical was clearly promising.[201]

In 1769 Kant introduced an explicit discussion of epigenesis in his course on metaphysics. Kant always used A. G. Baumgarten's *Metaphysica* as the basis for his course, and the topics concerning the soul ranged from discussions of human understanding to mind-body interaction and the afterlife.[202] In a section devoted to the origin of the soul, Baumgarten had rehearsed the reigning theories of organic generation: preexistence, spontaneous generation—Baumgarten's example here was infusoria—creation ex nihilo, and finally, "concreationism," according to which the soul was produced through some sort of transfer accomplished by the parents, a position derived from Aristotle's treatment of the matter. When preparing his own notes for this section, Kant wrote out the questions that would be addressed in his lecture: Was the soul a pure spirit before birth? Had it lived on the earth before? Did it live in two worlds—the pneumatic and the mechanical—at once? The questions were accompanied by a quick list of the various theories of generation, with Kant noting that the central division was between supernatural approaches to the question of origin and a naturalistic account, an account Kant described as an *"epigenesis psychologica"* (17: 416). The majority of Kant's commentary, however, was devoted to the comparative advantages of the preexistence theory of generation, in either its spermist or ovist variation, over the system proposed by epigenesis. In contrast to the preexistence theory, for example, the naturalistic system of epigenesis assumed material contributions from each of the parents, and this, Kant observed, required that prospective couples consider each other with greater care when planning to marry and reproduce.[203]

In later years, Kant would use this section of Baumgarten's text to discuss the properties of the soul and would invariably dismiss the possibility of its epigenesis.[204] In 1769, however, Kant's commentary focused on the physical aspect of generation, identifying epigenesis with a theory of blending that was in line with what he knew of Maupertuis's and Buffon's use of heredity as a basis for their arguments against preexistence theory. The next time Kant came to add notes to this section, epigenesis was again considered in terms of its biological claims, with Kant now explicitly linking the theory to the desired account of species generation he had first sketched in 1763. In his words, "The question is whether nature is formed organically (epigenesis), or only mechanically and chemically. It seems that nature does have spirit, given that in the generation of each individual there is a unity and connection of parts. And is there not also such a spirit, an animating essence, in animals and plants. In this vein one would have to assume an animating Spirit, operating within an original chaos, in order to explain differences between animals which can now only reproduce themselves" (17:591). This two-step model was the same proposed in Kant's *Only Possible Argument*, so far as an initially divine organization—out of an "original chaos"—was followed by the organic capacity for reproduction within the divinely delineated species lines. These two sets of comments, dated by Erich Adickes as having been written in 1769 and 1772–1776, respectively, demonstrate that during a period of crucial formation with respect to the development of Kant's system of transcendental idealism, Kant was actively aware of the epigenesis alternative to preexistence theories of generation.

More significant than Kant's commentary on Baumgarten for our purposes, however, is the set of notes Kant composed shortly after finishing his *Dissertation*. For in these notes, Kant explicitly connected theories of generation to systems of reason and to claims regarding the origin of ideas in particular. Distinguishing empiricists from rationalists, Kant identified his own position with the most radical possibility of all. As he sketched it, "Crusius explains the real principle of reason on the basis of the *systemate praeformationis* (from subjective *principiis*); Locke on the basis of *influx physico* like *Aristotele*; Plato and Malebranche, from *intuit intellectuali*; we, on the basis of *epigenesis* from the use of the natural laws of reason" (17:492). It was epigenesis, therefore, that Kant identified with the theory of "original acquisition" for explaining the generation of sensitive and intellectual concepts from the mind's own laws in the *Dissertation*. While it cannot be said for certain that Kant took epigenesis as his model when first drawing up his account of the origin of knowledge in 1770—though the evidence from 1769 certainly suggests this—it is certain that in the months following the *Dissertation*'s completion the connection had been made, that by then Kant had, to paraphrase Darwin, "at last got a theory by which to work."[205]

Concepts and Objects: Kant's Letter to Herz, 1772

Kant had presented his *Inaugural Dissertation*—with his former student Marcus Herz playing the role of disputant—on August 21, 1770. Twelve days later, Kant sent copies of the *Dissertation* off for feedback and, as with their earlier exchange, it was J. H. Lambert's response that

CHAPTER FOUR

would prove again to be of the greatest significance for Kant's developing project.[206] Kant had, in fact, never replied to Lambert's letter regarding the distinction between form and matter. "The reason," Kant now explained, "was none other than the striking importance of what I gleaned from that letter, and this occasioned the long postponement of a suitable answer" (10:96). Having dismissed the first and fourth sections of the *Dissertation* as discussions to "be scanned without careful consideration," Kant wanted Lambert's thoughts on the remainder of the work, for in Kant's own estimation of the remaining sections, "there seems to me to be material deserving more careful and extensive exposition" (10:98). Sections 1 and 4 of the *Dissertation* had covered topics that were traditional for metaphysics: the problem of intuiting the world as a whole versus as an aggregate and the difficulties in accounting for interaction between substances. The truly innovative work of the *Dissertation* appeared in the remaining parts of the text. Section 3 presented Kant's account of space and time as the originally acquired forms of sensible intuition. Section 2 laid out the strategy for certainty regarding empirical knowledge and introduced Kant's distinction between the laws at work in the "logical use" of the intellect and the "real use" by which pure concepts could be generated by attention to the working of these laws. The last section of the *Dissertation*, section 5, outlined Kant's method for metaphysics in light of the mind's susceptibility to surreptitious concepts or, as he had renamed these in the *Dissertation*, "subreptive axioms." These were the three sections meant for Lambert's inspection, and Kant summarized his general results for Lambert in a few lines:

> Space and time, and the axioms for considering all things under these conditions, are, with respect to empirical knowledge and all objects of sense, very real; they are actually the conditions of all appearances and all empirical judgments. But extremely mistaken conclusions emerge if we apply the basic concepts of sensibility to something that is not at all an object of sense.... It seems to me... that such a propaedeutic discipline, which would preserve metaphysics proper from any admixture of the sensible, could be made usefully explicit and evident without great strain. (10:98)

Kant had described his project as just such a "propaedeutic" (2:395) in his *Dissertation*, locating its success regarding the prevention of subreptive axioms in the radical break between sense and intellect.

Lambert replied within a matter of weeks, generously discussing the sections inquired after by Kant. From Lambert's perspective, the main challenges for Kant's theory lay in the ideality of space and time described in section 3 (this was the focus of Moses Mendelssohn's response to the *Dissertation* as well). But while Kant was willing to incorporate some of this in his later discussions of space and time (e.g., A36–37/B53–54), his general position regarding their transcendental ideality would not change. The importance of Lambert's letter for Kant lay rather in the remarks concerning section 2 of the *Dissertation*. Here Lambert was direct regarding what he saw as a problem facing the heterogeneity of sense and intellect as independent sources of knowledge: "My thoughts on this proposition have to do mainly with the question of *universality*, namely, to what extent these two ways of knowing are so *separated* that they *never* come together. If this is to be shown *a priori*, it must be deduced from the nature of the senses and of the understanding. But since we first have to become acquainted with these *a posteriori*, it will depend on the classification and enumeration of their objects" (10:105). Lambert's remark raised two concerns: first, the seeming impossibility of an a priori demonstration of sense and intellect's universal separation and, second, the need, as a consequence of that impossibility, to turn to their respective objects for evidence of their separation—a turn that would limit Kant to an a posteriori proof. Resorting to experience like this had been essential in Lambert's own ontological investigations, as he made clear further on in his reply: "It is also useful in ontology to take up concepts borrowed from appearance [*Schein*], since *the theory must finally be applied to phenomena again*. For that is also how the astronomer begins, with the phenomenon; deriving his theory of the construction of the world from phenomena, he applies it again to phenomena and their predictions in his *Ephemerides* [star calendar]" (10:108).[207] "In metaphysics, where the problem of appearance is so essential," Lambert advised, "the method of the astronomer will surely be the safest." For the metaphysician could also "take everything to be appearance, separate the empty from the real appearance, and draw true conclusions from the latter. If he is successful," Lambert concluded, then "he shall have few contradictions arising from the principles and win much favor" (10:108).

In the *Inaugural Dissertation* Kant had in fact emphasized the importance of distinguishing between empirical concepts garnered along the lines now suggested by Lambert and the "original acquisition" of pure concepts, concepts that, as he had put it, would "never enter into any sensual representations as parts of it, and could not, therefore, in any way be abstracted from it" (2:395). But Lambert had raised an important point nonetheless. How could one understand the fact that an intellectual concept like "cause," for example, should seem so readily applicable to experience and yet belong, by definition, to an entirely separate realm of knowledge? Sensitive knowledge, so neatly accounted for in the *Dissertation* via the forms of intuition and the processes of logical subordination, suddenly seemed deficient when explaining the experience of causal relations. Subreption had served as the catalyst for Kant's attempt to redefine metaphysics as a science of the limits and extent of human reason, but the radical separation of sense and intellect—the key to Kant's solution to the problem of logical subreption—might have to be rethought after all.

The problem was as follows: Subreption, as Kant initially conceived it, was unidirectional. It focused on the prevention of *sensible* concepts, concepts like "causality" and "force," being surreptitiously applied to objects of what would have to be a nonsensible intuition, objects like angels and souls. When Kant sat down to write the *Inaugural Dissertation*, however, his theory of cognition had outstripped the earlier conception of the problem. As Kant had explained in his first letter to Lambert, all he really had in 1765 was a negative account—a kind of "what not to do" for anyone interested in reconstructing metaphysics as a science of limits set by the boundaries of the human mind. By 1766's *Dreams of a Spirit-Seer*, this committed Kant to denying, for example, the possibility of direct knowledge of either human souls or the principles of life within nature. This prohibition automatically eliminated, therefore, the explicit objects to which sensible concepts were supposed to have been surreptitiously applied. The prohibition was carried over to the *Dissertation*, where, despite Kant's characterization of the break between sense and intellect as similar to the ancient distinction between the worlds of phenomena and noumena (2:393), there were in fact no noumenal objects to be found, and the intellectual intuition of such objects was flatly rejected (2:396). With noumenal objects thereby out of reach, subreption—still described as the central problem facing metaphysics—was reconceived as the result of misunderstanding the subjective nature of space and time as forms of human intuition.

According to the *Dissertation*, then, subreption appeared in three guises: it occurred when asserting that space and time could be applied to nonmaterial objects, as in attempting to spatially locate the soul within the body; it occurred when asserting that because we only experience objects in space and time, all objects are necessarily spatiotemporal, a fallacy in line with demanding that the universe have a beginning in time; and, finally, it occurred when asserting that intellectual concepts could *only* be applied to experience via space and time.²⁰⁸ This last example was a surprise, for here, in the closing moments of the *Dissertation*, Kant was discussing a case of intellectual concepts being applied to experience after having expressly forbidden it in section 2 (2:395). Was this a slip?

For the third type of subreption, Kant had taken his example from Crusius, who, according to Kant, illicitly filtered the intellectual concept of "existence" through the lens of temporality when declaring that "whatever exists contingently has at some time not existed" (2:417). Subreption occurred in this case by supposing that intellectual concepts required sensible intuition for their application. Crusius's "spurious principle," Kant explained, "arises from the poverty of the intellect, which for the most part discerns the nominal marks of contingency or necessity, seldom the real ones. Since, therefore, we can scarcely hope to determine, through marks derived *a priori*, whether the opposite of some substance is possible, we shall be able to do so only insofar as we have evidence that at one time the substance did not exist" (2:417). In the absence of any a priori discovery, in other words, the intellect turned to experience and, borrowing the concept of temporal change, illicitly declared it to be necessarily connected to the concept of contingency. But contingent existence, as Kant had already understood it in *The Only Possible Proof for the Existence of God* (1763), was also a way for seeing effects to be the result of God's free choice. Since God's activity was not susceptible to temporal laws, this marked the subreptive fallacy in connecting contingency and time. The argument seems to have distracted Kant from the fact that up until now in the *Dissertation* he had denied any possible connection between intellectual concepts and sensible experience, a denial motivated with respect more to maintaining the pure status of the intellectual concepts and moral exemplars than to the problem of subreption itself. In 1770, Kant still understood subreption to be a unidirectional problem, and the possibility that he might need to apply intellectual concepts to experience had simply not occurred to him.²⁰⁹

When Lambert questioned the universal separation of sense and intellect as independent modes of cognition, therefore, he might have pointed to Kant's discussion of Crusius, but his focus on section 2 pointed Kant back to his account of the intellectual concepts themselves. For here Lambert must surely have noted that "causality" was no longer considered a sensitive concept at all. Kant had indeed moved causality to his list of *intellectual* concepts, a decision undoubtedly reflecting the influence of Hume's skepticism regarding necessary connection. But while the a priori status of causality "protected" it from

Hume's skepticism, without a connection to sensible phenomena, Lambert seemed to suggest, metaphysics would remain not only sterile but ultimately useless in the face of the empiricist challenge.

In the wake of Lambert's response, it was clear to Kant that he would need to reconsider whether the separation of sense and intellect could be maintained at all. As he described this realization, "I noticed that I still lacked something essential, something that in my long metaphysical studies I, as well as others, had failed to pay attention to and that, in fact, constitutes the key to the whole secret of hitherto still obscure metaphysics" (10:130).[210] When Kant went on to describe this "key to the whole secret" of metaphysics to Marcus Herz, it turned on the problem of connecting intellect and sense. It was a problem of maintaining that "pure concepts of the understanding must not be abstracted from sense perceptions," that they "have their origin in the nature of the soul," and that "they are neither caused by the object nor bring the object itself into being," while *also* explaining how such pure concepts could be connected to objects at all (10:130). Kant could not have been clearer regarding the status of the concepts under consideration: their origin lay in the nature of the soul; they were neither abstracted from nor caused by the object; they were, in keeping with the *Dissertation*, original to the mind itself. This much had not changed. "And if such intellectual representations depend on our inner activity," Kant continued, "whence comes the agreement they are supposed to have with their objects—objects that are nevertheless not possibly produced thereby? And the axioms of pure reason concerning these objects—how do they agree with these objects since the agreement has not been reached with the aid of experience?" (10:131). This was the change. The trajectory of Kant's thinking since writing the *Dissertation*, so far as he now recounted it for Herz, turned on the problem of connection.[211] The problem of origin, by contrast, was no longer an issue.

Focusing on the problem of connection, then, Kant listed the kinds of relations that were easy to grasp. One could easily see, for example, how sensible content was connected to sensible representations, and it was also clear how an "archetypal" intellect could serve as the ground for its own representations. In mathematics, it was possible to understand how to connect axiom and intuition without experience because in this case "the objects before us are quantities only because it is possible for us to produce their representations," a production guaranteeing their connection (10:131). But the question of understanding how a concept like causality, for example, could both be generated a priori and yet conform to sensible experience "remained in a state of obscurity" (10:131).[212]

Moving on from the "obscurity" surrounding the problem of connecting sense and intellect, Kant proceeded to review theories regarding the origin of concepts. Since he had already listed the ease in understanding the relationship between sense data and sensible concepts, he now limited himself to theorists describing a priori concepts, since the locus of the problem of connecting them to sensible phenomena lay precisely in their purity. As Kant rehearsed the list, "Plato assumed a previous intuition of divinity as the primary source of the pure concepts of the understanding and of the first principles. Malebranche believed in a still-continuing perennial intuition of this primary being.... Crusius believed in certain implanted rules for the purpose of forming judgments and ready-made concepts that God implanted in the human soul just as they had to be in order to harmonize with things. Of these systems, one might call the former the Hyperphysical Influx Theory and the latter the Pre-established Intellectual Harmony Theory" (10:131). Kant dismissed such theories immediately, acidly noting that "the *deus ex machina* is the greatest absurdity one could hit upon in the determination of the origin and validity of our cognitions," a recourse encouraging "all sorts of wild notions and every pious and speculative brainstorm" (10:131).

The list Kant rehearsed was, of course, the same breakdown he had previously outlined for himself regarding theories of origin (minus the cases presented by Aristotle and Locke regarding empirical concepts): "Crusius explains the real principle of reason on the basis of the *systemate praeformationis* (from subjective *principiis*); *Locke* on the basis of *influxu physico* like *Aristotle's*; *Plato* and *Malebranche*, from *intuit intellectuali*; we, on the basis of *epigenesis* from the use of the natural laws of reason" (17:492). When employing biological vocabulary in his own notes, Crusius's belief in "implanted rules," for example, was identified with preformationism. In the letter to Herz, however, Kant was entirely focused on the question of connection, and the examples of theorists arguing for a nonempirical origin were therefore schematized in terms of their means for connecting a priori concepts and objects. The "Hyperphysical Influx Theory" defined systems where concepts and objects maintained connection because of their effective identity in God's mind. The "preformationist" theory maintained by Crusius relied on "Pre-established Intellectual Harmony" given God's work to establish all future potential connections between concepts and things at the moment of creation. For his own part, Kant was still clear regarding the epigenetic origin of concepts, concepts whose source lay "in the nature of the soul," but he had yet to discover a basis for connecting these to sensible objects.[213]

Kant was, however, ready to announce the progress he had made with respect to his understanding of the concepts themselves. Whereas the earlier list of concepts had been both short and somewhat vague—"possibility, necessity, substance, cause, etc. with their opposites or correlates" (2:395)—Kant now appeared to have in mind not only a specific number of concepts but, more importantly, a basis for their organization. And Kant took this advance to be important enough that he was ready to tell Herz that "so far as my essential purpose is concerned, I have succeeded, and I am now in a position to bring out a critique of pure reason that will deal with the nature of theoretical knowledge" (10:132). What precisely was this advance? Kant explained, "As I was searching in such ways for the sources of intellectual knowledge . . . I sought to reduce transcendental philosophy (that is to say, all the concepts belonging to completely pure reason) to a certain number of categories, but not like Aristotle, who, in his ten predicaments, placed them side by side as he found them in purely chance juxtaposition. On the contrary, I arranged them according to the way they classify themselves by their own nature, following a few fundamental laws of the understanding" (10:132). The origin of the intellectual concepts would no longer be generally based on the workings of the mind; they would from now on be indexed to particular mental laws as a means for their specific classification. If this was not yet to directly identify the intellectual concepts and the mind's laws for logical subordination, Kant was certainly very close to making this connection. This was significant, for it demonstrated that while Kant might still have been uncertain regarding the means for connecting a priori concepts and sensible objects, he was apparently close to adopting the successful model provided by sensitive knowledge when approaching intellectual cognition.

In the *Dissertation*, sensitive knowledge could be called "genuine knowledge" so long as the truth of a judgment of experience was determined by the *inner* coherence of the mental laws connecting subject and predicates in the judgment itself. As Kant put the point in 1771, "All truth consists in the correspondence of all thoughts with the laws of thinking and thus among one another" (17:524). Were Kant to embrace this model for cognition in general, to identify the laws of logical subordination with the intellectual concepts themselves, then the purity of the intellectual concepts would not be compromised, and the connection to sensible experience could be explained by their application to the *"apparentia"* delivered up by space and time. Embracing this model, however, would also mean accepting that the objects of cognition—objects dependent upon the mind in order for us to experience them—would henceforth be redefined as objects of knowledge. Nonetheless, it was precisely to this model, with these consequences, that Kant turned. And he did so in short order. By 1773 Kant summarized his position as follows:

> If certain concepts in us do not contain anything other than that by means of which all experiences are possible on our part, then they can be asserted *a priori* prior to experience and yet with complete validity for everything that may ever come before us. In that case, to be sure, they are not valid of things in general, but yet of everything that can ever be given to us through experience, because they contain conditions by means of which these experiences are possible. Such propositions would therefore contain the condition of the possibility not of things but of experience. (17:618)

This passage was Kant's response to the question he had just posed for himself: if "there are judgments whose validity seems to be established *a priori*, but which are nevertheless synthetic, e.g., everything that is alterable has a cause, whence does one arrive at these judgments?" (17:617). How do we achieve certainty, in other words, with respect to judgments that contain a synthesis of pure concept and sensible intuition? As Kant's response made clear, we achieve certainty by understanding such synthetic a priori judgments to be the means by which experience—so far as it is possible to *know* it, at least—becomes possible at all, a conclusion based, significantly, on the newly asserted identity between laws for logical subordination and concepts for conceptual determination. As Kant put it, "The concepts of the understanding express all the *actus* of the powers of the mind, insofar as representations are possible in accordance with their universal laws, and indeed their possibility *a priori*" (17:622). The "key to the secret of metaphysics," at least insofar as Kant had outlined the problem for Marcus Herz, appeared to have been found.

FIVE

From the Unity of Reason to the Unity of Race

The Unity of Reason

While preparing the preface for the *Critique of Pure Reason*'s appearance in 1781, Kant took time to describe the effort that had gone into his account of cognition, identifying the work done to connect concepts and objects, in particular, as that which had cost him "the greatest labor." Kant explained that there were in fact two parts to this discussion: "The one refers to the objects of pure understanding, and is intended to expound and render the objective validity of its *a priori* concepts. It is therefore essential to my purposes" (Axvi). This part of Kant's discussion, the so-called objective portion, turned on the problem and subsequent solution that we have seen in our discussion of Kant's letter to Herz from 1772. Indeed, Kant's introduction to the task of connecting concepts and objects in the *Critique* itself closely followed the outline of the problem as he had initially laid it out for Herz. In the first *Critique*, however, Kant moved quickly to a preliminary conclusion, arguing in terms close to those given in 1773 that "the representation is *a priori* determinant of the object, if it be the case that only through the representation is it possible to *know* anything *as an object*" (A92/B125). As for the second part of the work done to connect concepts and objects, Kant explained that this part "seeks to investigate the pure understanding itself, its possibility and the cognitive faculties upon which it rests; and so deals with it in its subjective aspect" (Axvi). Kant was immediately cautious regarding this piece of the discussion, emphasizing the independence of the two parts of his argument, and insisting that "the objective deduction with which I am here chiefly concerned retains its full force even if my subjective deduction should fail to produce that complete conviction for which I hope" (Axvii).[214] Now, since Kant had resolved, at least in outline, the problem of cognition with respect to objects by 1773, we have to assume that the greater part of his "greatest labor" had in fact turned on the problem of cognition with respect to the *subjective* portion of the discussion instead, the one seeking to understand the pure understanding itself.[215]

In Kant's notes *after* 1773, then, a thread can be picked up regarding a new set of reflections concerning the problem of unity. This was a topic that had already been broached by Kant with respect to space and time in the *Inaugural Dissertation*. There the main task had been to explain the manner in which the forms of intuition were not only a part of sensible experience but indeed served as the ground of its universal interconnection as well (2:398). The expositional problems regarding this task had proved especially challenging for Kant. Kant needed to show that space and time were distinguishable as forms or grounds of experience, even as the experience of a continuous spatiotemporality was their special yield. Complicating this narrative was his sense that the specific work of "temporalizing" sense data could itself be seen as a process taking place in time (2:400). Time could be seen as a part, therefore, of all three of the logically distinct moments of synthesis: it stood outside of the process as a form of intuition, it seemed to be itself in time as it worked to turn an aggregate of sensible impressions into a successive series of spatiotemporal intuitions, and it was of course an indelible part of the resulting experience, given that, from the subject's perspective, all sensible phenomena were locatable in time. Adding to Kant's expositional difficulties regarding all this was his insistence that space and time were themselves, as a priori forms of intuition, originally generated through the activity of the mind (2:401, 403). Such narrative problems aside, however, the main point remained the same: space and time could serve as reliable grounds for the coherent unity of experience only because they were themselves already unities. And it was this feature, more than anything else for Kant, that made them a more suitable basis for experience than the models proposed by either Leibniz or Newton (2:403).

In the mid-1770s Kant seems to have returned once again to his 1770 discussion of sensitive knowledge, therefore, when reflecting on the need to further account for grounds supporting the unity of experience. Since the *Dissertation* Kant had continued to deepen his understanding of the logical processes at work in conceptual determination. These were processes now described in terms of the mind's "disposition" (*disposition*) or "aptitude" (*aptitudo*) (17:655, 656, 660, 662) for the organization of sense data according to rules, an organization yielding the "exposition" (*exposition*) (17:643, 660, 661) of the rule in the form of representations. As Kant put it, "The exposition of appearances [*Erscheinungen*] is thus the determination of the ground on which the interconnection of the sensations in them depends" (17:643). But Kant also began at this point to address the need to logically distinguish between a comprehensive ground for unity—"The condition of all apperception is the unity of the thinking subject" (17:651)—and the particular unities of intuition and the conceptual rules that were together responsible for generating coherent representations.[216] Thus, in his words, "The unity of apprehension is necessarily combined with the unity of the intuition of space and time, for without this the latter would yield no real representation," but ultimately the "principles of exposition must be determined on the one side by the laws of apprehension, on the other by the unity of the faculty of understanding" (17:660). Distinguishing in this manner between the unity of experience and the unity of reason (17:709), Kant was ready to insist that "there must be principles of the self-determination of reason, which are different from those in which reason is determined by appearances and their conditions. These are principles of the unity of cognition as a whole, hence not of partial but of total unity" (17:711).[217]

Kant did not yet identify what the special principles underlying reason's self-determination might be. He had forbidden positive discussion of such self-determination in the case of the organism, insofar as this seemed to require appeals to a soul or "principle of life." In his notes regarding the unity of reason, however, Kant flirted with just this possibility:

The understanding itself (a being, that has understanding) is simple. It is substance. It is transcendentally free. It is affected by sensibility (space), it is in communion with others. . . . Everything is grounded on an original understanding, which is the all-sufficient ground of the world. The necessary unity of time and space is transformed into the necessary unity of a primordial being, the immeasurableness of the former into the all-sufficiency of the latter. The beginning of the world in time into its origin. The divisibility of appearances into the simple. (17:707)

The *Critique of Pure Reason* would later dismiss such notions regarding the understanding as "simple" or "substance," as cases of "transcendental illusion"—the new name for the fallacy of logical subreption (A348–361)—and the overtones of the understanding as the "all-sufficient ground of the world" would for the most part be gone. If there were special principles for the self-determination of reason, therefore, they lay elsewhere than in reason's explicit identification with the soul. In the midst of these considerations Kant put the work aside, however, taking the time instead to produce an essay that would serve as an advertisement for his upcoming course on physical geography. By 1775 it had been four years since Kant had published anything; the advertisement would surely be noted.

The Unity of Race

The period from 1770 to 1781 is almost invariably referred to as Kant's "silent decade." During these years Kant published only rarely and nothing that would seem obviously to concern the forthcoming *Critique*: a brief review of a book by an Italian anatomist named Pietro Moscati (2:421–425, 1771), a course announcement (2:427–443, 1775, rev. 1777), and two short pieces supporting Johann Basedow's Philanthropinum school, a school dedicated to the principles of education as had been outlined by Locke in *Some Thoughts Concerning Education* (2:445–452, 1776, 1777).[218]

Leaving the Moscati review and Kant's support for Basedow's school aside for the moment, it is worth considering Kant's motivation in advertising a course at this juncture of his career at all. Kant had advertised courses four times before. The first three times (1757, 1758, 1759-1760) he had included essays connected to the announcement of his lectures on physical geography, and between the novelty of the course and Kant's need as a new instructor to generate a paying audience, the three early advertisements made sense. Kant was quick to develop his reputation as a successful lecturer, however, and the courses on physical geography were extremely popular from the start, so there were different grounds for his next course announcement. This appeared in 1765–1766, and the special focus of this essay was the proper method of philosophy, a topic chosen in the wake of Kant's growing suspicions regarding the pervasive use of "surreptitious concepts" in metaphysics. It was now ten years later, and Kant had already held the

CHAPTER FIVE

long-sought-after chair in logic and metaphysics for five years. The course on physical geography had also matured. By 1765–1766 Kant had begun to compress the portions of the course devoted to "physical, moral, and political geography"—subjects Kant took to be directly connected to the physical features of the earth—in order to make room for his increasingly expanded discussions of human nature, or *"man, throughout the world, from the point of view of the variety of his natural properties and the differences in that feature of man which is moral in character"* (2:312). Indeed, by 1772–1773 the course had become so full that Kant began to offer a separate course on anthropology, subsequently alternating the two courses between the university's summer and winter semesters for the remainder of his career. But while this might have suggested a basis for an announcement regarding the new course devoted to anthropology, Kant chose instead to advertise the course he had taught some eighteen times before. In 1765 Kant knew that he had discovered something important when diagnosing subreption as the source of the general disrepute into which metaphysics had fallen; it was a discovery worth announcing. What had Kant discovered in 1775 that he felt it necessary to announce?

Since 1757 Kant had followed Buffon's model when teaching physical geography, and just as Buffon had devoted lengthy rehearsals to what he termed "The Varieties of the Human Species," Kant had as well. For Buffon, "these varieties may be reduced to three heads: 1. The colour; 2. The figure and stature; and, 3. The Dispositions of different people."[219] In Kant's 1757 announcement of the course, he too explained that he would be comparing people from different regions in terms of their color (*Farbe*), their natural shape (*natürliche Bildung*), and their dispositions (*die Neigungen*) (2:9). Kant agreed with Buffon, moreover, that food (*Landesprodukte*) and climate (*himmelstriche*), alongside "manners," as Buffon had also described dispositions, were the central contributing factors to the many obvious differences existing between people around the world.

The discussion of race for Buffon, and Maupertuis before him, was intimately connected to an account of generation insofar as differences or varieties were supposed to trace their origin to unspecified processes during embryogenesis. Racial difference was thus discussed by the French theorists in terms of biological considerations of "blending" and inheritance ahead of any taxonomical considerations regarding specific divisions between species, varieties, and races. Maupertuis, for example, had been long conscious of the difficulties facing attempts to discover the principles of inheritance, and his documentation of the reappearances of traits such as polydactylity in the Ruhe family had been used in support of his own account of generation against preexistence theorists, even as it served as a source of frustration given his inability to discern its underlying laws. From experiments upon his own menagerie, Maupertuis was well aware, moreover, of the special difficulties in creating varieties (by crossing) that would consistently breed true and not simply revert to type within a few generations. Without the means for investigating further the mechanisms for the inheritance of skin color as a particular trait, therefore, Maupertuis concentrated instead on an account of its origin.

Observing that variations in color are no more noteworthy from nature's perspective than any of her other variations, Maupertuis took breeders to hold much the same view. "Blackness," as he put it, "is just as inherent in crows and blackbirds as it is in Negroes, but I have often seen white crows and white blackbirds. Such varieties would undoubtedly become breeds if they were cultivated." Maupertuis did not consider whether white negroes could be similarly cultivated, but he took the appearance of albinos within black African families to be akin to the appearance of white crows, and he took each of these phenomena to be clues for discovering the origin of black skin.[220] For what such rare appearances demonstrated, according to Maupertuis, was the element of chance during the relatively plastic processes of embryogenesis: "Nature," he explained, "holds the source of all these varieties, but chance or art sets them going."[221] Given Maupertuis's particulate theory of generation, all variations would have to be traceable to the seminal seeds and their specific organization in the formation of an organism. As he put it, given that it is "the matter of the seminal fluid of each animal from which parts resembling it are to be formed," it "might not be unlikely to suggest that each part furnishes its germs [*germes*]."[222] Variation of any kind, therefore, was due to chance events during formation, and the maintenance of the variation across generations was taken to be the result of art, a fact continually demonstrated by the work of breeders. Speculating on the relative frequency of "white children" (albinos) produced by black parents in comparison to black children from white parents, Maupertuis argued that there must be a greater abundance of white particles in the ancestral stock of the species. By chance, black skin had appeared as a trait; by art—at least presumably, since Maupertuis did not suggest an explanation for the relative stability of this trait compared to others—the trait had become heritable as part of a distinctive variety. Less interested in accounting for geographic distribution, Maupertuis considered the effects of climate and food only in passing.

In his words, "Though I imagine the basic stock of all these varieties is to be found in the seminal fluids themselves, I do not exclude the possible influence of climate and food. It would seem that the heat of the Torrid Zone is more favourable to the particles that compose black skin than to those that make up white skin. And I simply do not know how far this kind of influence of climate may go after many centuries."[223] For Maupertuis, this theory of generation was sufficient to account for the origin of differences in skin color, even if it failed to account for any specific principles regarding its inheritance and relative stability across generations.

Like Maupertuis, Buffon also located the source of variations in the processes of embryogenesis. He advanced an explanation of both heritability and geographical distribution, however, insofar as he located variation, or "degeneration," from the original stock as he later put it, in the capacity of organic molecules to affect the internal molds of an organism. So far as these molecules were initially in the soil, all initial variation, according to Buffon, was the direct result of food. "Transplantings," migrations, these actions led species to change the organic bases of their phyletic lines, since climate and soil combined to affect qualitative changes between even identical foods, between grasses grown in the plains, for example, and those same grasses grown at higher altitudes.[224] Once the change had been made to a species' internal molds, a degenerate form such as the common sheep, for example, would continue to produce sheep instead of a "mouflon," which Buffon took to be the original formation or mold for the species line. When considering the varieties of men, Buffon took color to be a superficial variation compared to actual differences in shape.[225] Whereas climate alone might account for the effect of color, food was required in order to effect the internal molds, such that the species could take on observable differences in stature and proportion. As he summarized it,

> It is chiefly by aliment that man receives the influence of the soil which he inhabits: that of air and climate acts more superficially. While the climate changes the colour of the skin, food acts upon the internal form by its qualities, which are always related to those of the earth by which it is produced. . . . Hence, in countries remote from the original climate, where the herbs, fruits, grains, and the flesh of animals differ both in quality and substance, the men who feed upon these articles must undergo still greater changes.[226]

This, in short, was the basis of Buffon's explanation of differences in color, shape, and even disposition between people. All humans, however, were members of the same species, according to Buffon, for "even apart from the bible's instruction regarding Adam," humans of all shapes and dispositions were "capable of uniting, and propagating the great and undivided family of the human kind."[227] It was indeed a tribute to mankind's special capacities—its "greater strength, extension, and flexibility"—that as a species it had been able to spread out and flourish despite all manner of differences in soil and climate in the world. For Buffon, such adaptability lay not in the physical attributes of the species but rather "more on the qualities of the mind than those of the body. . . . By the powers of genius he [man] supplied all the qualities which are wanting in matter."[228] Genius, then, explained the geographic distribution of mankind; the results of that distribution explained humanity's manifold variations: "The blood is different," Buffon declared, "but the germ is the same."[229]

By the time Kant sat down to compose his own account of the origin of racial differences, he had long been aware of the integral connection between the theories of embryonic generation espoused by Maupertuis and Buffon and their explanation of the origin of racial characteristics. Indeed, Kant's earliest consideration of epigenesis as a particulate theory of generation in comparison to individual preformation theories—with the French theorists as his presumed models of the former—had emphasized the added difficulty of having to consider the possibility of racial blending during embryogenesis, a fact that led Kant to consider preexistence theory as at least more practical as a theory when it came to choosing a spouse.[230]

Kant's own approach to the question of race would be different. And it would have to be, given that Kant's attitude toward subreption was as applicable to investigations in the life sciences as it was to dogmatic metaphysics: each sought to make impossible claims regarding the special properties of the "principle of life" in matter, in the one case, and the human soul in the other.[231] The French theorists might have been comfortable advancing such speculations as the basis for an explanation of race, but Kant could hardly promote his own position as the result of any theory of biological generation. On the contrary, then, in Kant's essay he would declare himself to be merely advancing an "idea" intended for "useful academic instruction," a mere preparatory exercise contributing to an enlarged "pragmatic knowledge of the world" (2:443). By advancing his investigation under the framework of an idea meant solely for rational consideration, Kant was adopting a new methodological stance, one capable of philosophical speculation into the forbidden territory of biological origins, for example, while yet avoid-

CHAPTER FIVE

ing the epistemic pitfalls of subreption.[232] It was this new methodological attitude that marked the discovery worth announcing in 1775.

In much the same manner as Buffon had earlier attempted to replace Linnaeus in questions of natural history, in Kant's discussion of race he was determined to replace them both. Describing Linnaeus's "school system" as a system concerned with merely the "description of nature" (2:434), Kant argued that such an approach sought only to divide classes according to resemblance and thereby "bring creatures under titles" (2:429). This was not natural history; it yielded only a "school system for memory," with the obvious reference being to the old taxonomy of the *Encyclopédie*, according to which history and natural history alike belonged to the province of "Memory." In the *Encyclopédie*, discussions of zoology and botany—that is, investigations into the sources of animal and plant physiology, into their *functions* as opposed to their *forms*—were included in the general category of physics and were thus classified under the province of "Reason." It was to this area that Kant seems to have assigned Buffon. For against the artificial divisions by resemblance proposed by Linnaeus, Kant cited "Buffon's rule" regarding interfertility as a basis for the "natural division" of nature into species and kinds; what Buffon offered was a division based on reproduction, an animal's basic function, versus the mere identity of forms. By locating phyletic lines (*Stämme*) according to the rule for generation (*Erzeugung*), Kant credited Buffon with having provided a "natural system for the understanding" that could thereby "bring creatures under laws" (2:429) as opposed to mere titles. But while Buffon might have offered a better or more natural means for ordering species, his account had not achieved the status of a genuine natural history, according to Kant, since he took Buffon's results to remain ultimately taxonomical, versus historical, in his consideration of species. As Kant described the state of investigations into natural history,

> It is clear that the cognition of things as they *are now* always leaves us desirous of the cognition of that which they once *were* and of the series of changes they underwent to arrive at each place in their present state. *Natural history*, which we still lack almost entirely, would teach us about the changes in the shape of the earth, likewise that of its creatures (plants and animals) that they have undergone through natural migrations and the resultant subspecies from the prototype of the phyletic species. It would presumably trace a great many of seemingly different kinds to races of the same species and would transform the school system of the description of nature, which is now so extensive, into a physical system for the understanding. (2:434)

Only the genealogical approach to species as a temporal whole across time, in other words, could count as natural history. What Kant wanted was a history yielding neither the artificial nor the natural systems of Linnaeus's or Buffon's taxonomical considerations but one that could indeed produce a "physical system for the understanding." Of course it was precisely this kind of genealogical-historical approach that had led Buffon to effectively weaken his own "rule" regarding conspecificity in his late article "Degeneration" (1765). The search for empirical evidence of fertile mules dominated Buffon's discussion in that essay, since only these could allow him to describe seemingly disparate species as in fact varieties of older, degenerated lines. Buffon's aims, at least, were therefore perfectly in line with Kant's definition of genuine natural history. Buffon's mistake, from Kant's perspective, was concentrating on a physiological explanation of the origin, degeneration, and even potential reversion of varieties; it was the *subreptive* means by which Buffon sought to unify nature, not the goal itself, that was the problem for Kant.

Having thus dispatched Linnaeus and Buffon to their respective locations under "Memory" and "Reason," Kant took up the problem anew, starting with a dizzying array of taxonomical terms intended to discover the precise definition of "race."[233] Kant took races (*Racen*) and varieties (*Varietäten*) to each represent "subspecies" (*Abartungen*), with the point of distinction being the persistency with which characteristics were inherited: varieties were inconsistent, but races were not. Kant accepted Buffon's definition of degeneration (*Ausartung*) as a case when "subspecies could no longer provide the original formation of the phylum [*Stammbildung*]" (2:429) before moving on to further distinguish "strains" (*Spielarten*) from a "sort" (*Schlag*). What made the category of race stand out in the taxonomy was the special capacity of racial character to withstand complete amelioration through either racial blending (*Mischung*) or transplanting. Because Kant followed Buffon in assigning stature and disposition to contingent geographical forces like climate and nutrition, however, he took color—which Maupertuis and Buffon had each dismissed as only a superficial mark of distinction—to be the unique and permanent identifier of race. This meant that while "sorts" of races could be distinguished by their figure as a result of their geographical location, their belonging to a particular race would remain stable, even were the sort to eventually change shape again due to transplanting. As Kant explained it, "The condition of the soil (humidity or aridity), likewise that of nutrition, gradually introduce a hereditary difference or *sort* among animals of the same phylum and race,

CHAPTER FIVE

chiefly with respect to size, proportion of the limbs (heavy or thin), as well as natural disposition [*Naturells*], which, while resulting in half-breeds in mixing with foreign ones, disappears over the course of few generations on other soil and with different nutrition (even without a change of climate)" (2:431). Kant understood that it was important to retain this aspect of Buffon's account regarding geographic distribution, since it explained the different appearances or "sorts" of people across regions that seemed to offer similar conditions. Because soil and nutrition were responsible for the changeable aspects of racial sort, climate became the special "occasioning cause" (*gelegentliche Ursache*) of nonchangeable differences in color (2:436).[234] But while Kant understood all of these causes to work their effects only so far as they took root in a "generative power" (*Zeugungskraft*) during embryogenesis—a necessary requirement for the subsequent inheritance of traits (2:436)—he was careful to base his account on something other than an empirical account of animal generation.

This meant striking a different path than that taken by either Maupertuis or Buffon. In the 1763 essay in which Kant had first criticized Maupertuis's and Buffon's accounts of generation as being either "incomprehensible" or "entirely arbitrary inventions" (2:115), Kant had in fact also anticipated to some degree the strategy he would now take.[235] In the older essay Kant had been equally dissatisfied with preexistence theory and the systems of Maupertuis and Buffon, so far as each seemed to prevent the possibility of genuine activity on the part of the organism. Thus despite Buffon's protestations to the contrary, his supposed "natural order of unfolding [*Auswickelung*]," as Kant saw it, was not a "rule of the fruitfulness of nature, but a futile method of evading the issue."[236] What Kant had wanted was an alternative to these positions, one that "granted to the initial divine organization of plants and animals a capacity, not merely to develop [*entwickeln*] their kind thereafter in accordance with a natural law, but truly to generate [*erzeugen*] their kind" (2:115). As Kant went on to consider what he took to be better grounds for a revised physicotheology, he focused on God as the only logical basis of a necessary unity in nature and as the sole ground, therefore, upon which a subsequent manifold of contingent "forms of adaptedness [*Tauglichkeit*]" could be orchestrated. This capacity for environmentally contingent adaptation on the basis of an originally divine organization, according to Kant, was evident throughout nature as a whole. As he described it, "One will presume that the necessary unity to be found in nature is greater than strikes the eye. And that presumption will be made not only in the case of inorganic nature, but in the case of organic nature as well. For even in the case of the structure of an animal, it can be assumed that there is a single disposition [*Anlage*], which has the fruitful adaptedness to produce many advantageous consequences" (2:126).[237] Without such a disposition, Kant remarked, one would have to suppose all manner of "special provisions" to produce such effects. From Kant's perspective it made better sense to avoid needless multiplication of causes and to generally reduce the tincture of the supernatural—special provisions and miraculous events—as much as possible when discussing nature. It was much more fruitful to suppose instead the active workings of organic life—generation (*Erzeugung*) versus unfolding (*Auswickelung*)—even while remaining mindful that the only plausible means for asserting that nature was both freely adaptive yet necessary in its ultimate unity of effects, was through an appeal to some kind of supersensible ground.

In the years since these reflections on a revised physicotheology, Kant had consigned such proofs for the existence of God to the realm of dogmatic metaphysics, but the spirit of the earlier discussion would remain. For rather than a direct appeal to God as the basis of unity in nature, Kant now identified "Nature" itself as an unspecified ground of concern with respect to the deeply interconnected flourishing of her creatures. Maupertuis might have been right, therefore, to emphasize the role of chance in the appearance of varieties already contained "in the basic stock" of the seminal fluids, but the fact that such varieties were possible at all, according to Kant, was ordained by nature as a matter of necessity from the start. "This care of Nature," Kant announced, "to equip her creature through hidden inner provisions for all kinds of future circumstances, so that it may preserve itself and be suited to the difference of the climate or the soil, is admirable" (2:434).

As in the essay from 1763, the "inner provision" for adaptation in the 1775 essay on race was assigned to an organism's disposition—now called "natural" (*natürliche Anlage*)—for changes in size and proportion, with a description of germs (*Keime*) added by Kant in 1775 for the production of new parts. Thus birds transplanted to colder climates would have germs ready to be unfolded (*ausgewickelt*) for the development (*Entwickelung*) of new parts, that is, more feathers. And wheat faced with cold could rely on the unfolding of a natural disposition regarding the proportional thickness of its protective chaff (2:434). The vagaries of environment thus served as contingent occasioning causes for changes in the creature, but the grounds for an individual's adaptive response were prepared in advance because of nature's concern for the species lines under her protection. Such advance concern introduced the language of purpose (*Zweckmäßigkeit*) and ends (*Zwecke*) into Kant's discussion, vocabulary that could only be employed insofar as what was being advanced was an "idea" meant to aid in our investigation of the world. "Chance or the universal mechanical laws could not produce such agreements," Kant argued. "Therefore we must consider such occasional unfoldings as *preformed* [*vorgebildet*]" (2:435). Preformed germs and dispositions were thus purposed from the start for their later formation into traits meant to allow a species' adaptation to its environment. The great adaptability of mankind meant that the species' widespread geographic distribution was a matter of destiny: "The human being was destined for all climates and for every soil" Kant wrote, and

> consequently various germs and natural predispositions had to lie ready in him to be on occasion either unfolded or restrained, so that he would become suited to his place in the world and over the course of the generations would appear to be as it were native to and made for that place. With these concepts, let us go through the whole human species on the wide earth and adduce purposive causes of its subspecies therein in cases where the natural causes are not easily recognizable and again adduce natural causes where we do not perceive ends. (2:435)

Armed thereby with an approach to natural history that combined purposive and natural causes, Kant was free to assert on teleological grounds not only the *historical* unity of the species in the case of mankind—a unity happily supported by the empirical experience of interfertility between the races—but a noncontingent basis for the subsequent appearances of traits serving to differentiate individuals from the other members of the species. In this sense, therefore, Kant could have said very well with Buffon that "the blood is different, but the germ is the same."[238]

A Germ of Reason and a Germ for Race

The language of germs and dispositions purposed for ends special to the development of mankind appeared in Kant's other published writings during the 1770s as well.[239] In his earliest piece, the short review of Moscati's anatomy textbook in 1771, Kant had concurred with the physiologist regarding our animal setup, but he had also removed mankind's "rational nature" from such material considerations. Here the language of teleology was already at work, for while "the first foresight of nature" might have been oriented toward the human being as an animal, Kant argued, there was also "placed in him a germ of reason through which, if the latter develops, he is destined *for society*, and by means of which he assumes permanently the most suitable position for society, viz., the *two-footed* one" (2:425). This posture, pace Moscati, might be against our animal nature, but man could surely "live with the discomforts which result for him from the fact that he has raised his head so proudly above his old comrades" (2:425). This theme would be picked up again in the years to follow, most immediately in Kant's lectures on anthropology, where discussion of man's perfectibility was tied to germs for reason, for good, and for character. As Kant put it at one point during the 1775–1776 winter semester, "Innate to human nature are germs which develop and can achieve the perfections for which they are determined" (25:694, cf. 15:500).[240]

In the two endorsements Kant published on behalf of Basedow's Philanthropinum school in 1776 and 1777, the message regarding nature's plan for mankind was repeated, in this case grafted to organic imagery by Kant's employment of botanical metaphors throughout his endorsement of the school.[241] In these essays Kant explicitly likened the school to a plant, a creature that like any species was determined to survive through the dispersal of its seeds (*Samen*) but whose germ (*Keim*) required protection and care while still young (2:448). Like any other natural organism, the school-cum-plant fell under the general offices of "Nature herself," and Kant took it to thereby face a set of particular demands regarding its place in the economy of nature. Its first task as an organism concerned the preservation of itself as a species through either propagation or the dispersal of its seeds. Kant described this self-preservation through reproduction in terms of both the founding of additional schools and the formation (*Bildung*) of well-instructed teachers (2:449). Here Kant played on the idea of cultivation, moving between images of the school as a site of organic generation, as itself a "nursery" (*Pflanzenschule*) capable of producing teachers as its particular cultivars, and as an actual plant, one capable of its own organic generation and functioning thus as "a seed which, by means of careful cultivation, can give rise in a short time to a multitude of well-instructed teachers who will soon cover an entire country with good schools" (2:450).

The second task nature had given to the school concerned its role in the support and cultivation of mankind, nature's most favored phyletic line. Regarding this task, Kant turned again to the idea of cultivation. The soil or ground upon which Basedow's methods were to take effect, according to Kant, had been prepared by nature in advance, given that they lay within man as his natural predisposition to such cultivation (*Ausbildung, Bildung*). The school's cultivation of natural predispositions and the ability of men to thereby become cultivated members of society were what it meant to talk of the preservation and advance of the human species for Kant. "For," Kant explained, "that which is merely the development [*Entwickelung*] of the natural predispositions [*natürlichen Anlagen*] lying in humanity shares this feature with universal mother nature: that she does not allow her seeds [*Samen*] to run out, but rather multiplies herself and preserves her species" (2:447). By promoting the development or advancement of mankind in this manner, Basedow's was an institute that was therefore as "fitting to the purposes of nature" as it was to the purposes of society (2:447). The school was thus capable of "the greatest possible, most permanent and universal good": it served as a site "where the seed [*Same*] of the good itself can be cultivated and sustained, in order that in the course of time it may disseminate and perpetuate itself" (2:451). In light of this, Kant urged his readers to "cultivate with care this still tender germ [*Keim*]" of a school, in the hope that it might achieve "complete growth" and have "its fruits soon spread to all countries and to the most remote descendents" (2:448).[242]

The discovery worth announcing in 1775—which can be recognized as the clear backdrop for Kant's interest in the Philanthropinum school—was thus an increasing sense on Kant's part of the positive explanatory role that could be played by teleology in the search for a rationally unified order, for something that was at work in the nature of the human being as much as it was in "Nature herself." Oriented by the language of destiny and purpose, Kant discovered that with a teleological approach he could avoid the pitfalls of subreption, even while invoking the beneficence of nature's care for her chosen species. The ends of nature and humanity could be connected, even identified, moreover, once the grounds for their unity could be located outside the push and pull of empirical experience. Nature had provided mankind with a germ of reason and with dispositions intended for the gradual perfection of the species as a whole.[243] She had worked, at the same time, on man's physical nature, urging him toward every corner of the earth and thereby transforming him until he became original to and indeed a product of the land on which he dwelled. The natural processes of distribution, adaptation, and inheritance could be understood once they were reframed as the intentional work of a nature striving to develop "the sleeping powers of humanity" (2:431), for the unity of Nature's intentions demanded not only the underlying unity of the species but a divergence of its traits as well. Kant's argument from experience to its grounds thus led past both chance and mechanism when accounting for divergence, tracing itself back to an idea of natural providence as alone capable of grounding the unity and difference of humankind.

These considerations were intimately connected to Kant's work to develop a theory of human cognition. The preoccupations in the intellectual sphere were the same: the need to identify grounds for unity in reason and experience, and the role assigned to predispositions for organizing an inherently contingent sensible content. There was, however, a crucial difference when it came to the consideration of mental grounds and dispositions and the idea of germs meant to account for the adaptive potential of the human species. The fact of human freedom, according to Kant, meant that the basis of our particular cognitive unity had to be generated by us. Only the slightest tincture of supernatural concern could be allowed, therefore, if Kant was to maintain a balance between something that was genuinely free, indeed author of itself, yet susceptible to orientation. As Kant described this in the *Critique of Practical Reason*, the task was to understand a human being as a being with "an aptitude for purposes generally, i.e., in a way that leaves that being free" (5:431).[244] It was this task that explained Kant's appeal to a germ of reason, a germ that was not preformed in the sense of something implanted, but predisposed in the manner of an innate orientation, an orientation grounding the aptitude for an ordered experience as much as it did the cultivation of an ordered society. The mind was innately predisposed toward the organization of sensible content, therefore, but the means for this organization—the originally generated concepts and rules—would have to be generated. Kant would make the same point in 1790, arguing that "the ground of sensible intuition" is a "mere *receptivity* peculiar to the mind, when it is affected by something (in sensation), to receive a representation in accordance with its subjective constitution. Only this first formal ground, e.g., of the possibility of an intuition in space, is innate, not the spatial representation itself. . . . Thus arises the formal *intuition* called space, as an originally acquired representation (the form of outer objects in general), the ground of which (as mere receptivity) is nevertheless innate" (8:222).[245] The innate "laws of the mind," appealed to as the basis for the original generation of space in the *Inaugural Dissertation*, were replaced by Kant during the 1770s with a notion of receptivity that was no less innate but far less substantial. For such receptivity, like all such mental predispositions, indicated nothing more for Kant than a set of

CHAPTER FIVE

possibilities whose actualization was by no means determined. As he continued his remarks from 1790 in this vein, this was the case, moreover, regarding the epigenesis of intellectual concepts as well. An empirical concept, he explained, was always derivative so far as its origin was concerned (*acquisitio derivativa*), for "it already presupposes universal transcendental concepts of the understanding, which are likewise acquired and not innate, though their *acquisitio*, like that of space, is no less *originaria* and presupposes nothing innate save the subjective conditions of the spontaneity of thought (in conformity with the unity of apperception)" (8:223). Kant held that the unity of apperception, as ground of the unity of experience, stood apart from the material conditions it would itself come to establish so far as their experience was concerned. This was part of what it meant to identify apperception with the spontaneity of thought. The only thing innate to the mind, therefore, was its deep sense of possibility, of the mind as a site of spontaneity and freedom, of freedom that could be perfected or realized in the creation of itself and its experience through the act of cognition. As for the laws of the mind, the constraints or means for the realization of cognition, these were now understood to be emergent, arising from reason itself as part of its own subjective conditions, as part of its predisposition to unity.

In the piece from 1790, *On a Discovery Whereby Any New Critique of Pure Reason Is Made Superfluous by an Older One*, Kant had been called upon to defend his position against claims made by Johann Eberhard and others regarding the supposedly deep similarities between the new transcendental account of cognition and the older metaphysical one proposed by Leibniz.[246] It was in the midst of his defense that Kant had sought to define more precisely what he considered to be innate to the mind. It is worth recalling once more, therefore, Leibniz's own description, according to which the mind existed as a thing predisposed to the discovery of innate truths within it: "It is not a bare faculty consisting in a mere possibility of understanding those truths," Leibniz wrote, "it is rather a disposition, an aptitude, a preformation, which determines our soul and brings it about that they are derivable from it."[247] Now it can be seen that Kant too considered the mind to be something innately predisposed, but in this case it was a predisposition for the active generation, as opposed to the mere unfolding, of cognition and its objects alike.[248]

When Kant returned to work on the *Critique of Pure Reason* in the years following his first essay on race, he picked up exactly where he had left off in locating the mind's generation of rules original to itself as the special characteristic of critical philosophy. Kant's approach at this point, as in so many of his notes regarding this, was both visually and conceptually taxonomical. Intellectual concepts could be considered as either preformed *"educta"* or epigenetic *"producta"*; as products, the generated concepts could be considered a posteriori, if they were acquired from physical experience, or a priori, as Kant had it, if they were instead "occasioned by experience through our awareness of the formal characteristics of our sensibility and understanding" (18:8). With his account of space and time as the a priori forms of sensible intuition in mind, Kant noted that the real point of distinction between concepts was therefore not whether they were sensitive or intellectual but rather whether they were acquired a priori or, as in Aristotle and Locke, a posteriori. As usual, Kant grouped Plato with Leibniz so far as each offered a theory of a priori intellectual cognition, but in contrast to Kant's account of the original acquisition or generation of a priori concepts, their theories were based on divinely produced ideas, ideas that could only be accessed via "mystical intuition" in the case of Plato or by "intellectual intuition" in the case of Leibniz. Kant's own account of "transcendental concepts," by contrast, was oriented toward their discursive use—"our intuition is physical not mystical; the physical [is] organic not pneumatic" (18:13)—and their origin, as he repeated the point, was *"per epigenesin intellectualem"* (18:12). Only "intellectual epigenesis," as the basis of Kant's account of cognition, could adequately respond to the challenges faced by metaphysics, since only an account securing the origin of knowledge in this manner could respond to the challenge of skepticism without thereby losing sight of human freedom as one of the central objects of metaphysics itself.

SIX

Empirical Psychology in Tetens and Kant

Epigenesis and Evolution in Tetens's *Philosophical Essays*

For those interested in reconstructing Kant's path toward the *Critique of Pure Reason*, it has become standard practice at this point in the reconstruction to mention the absence of any evidence, in either Kant's notes or in his lectures, of the important role that would be later assigned to the transcendental imagination. Once this absence has been noted, J. N. Tetens's *Philosophical Essays on Human Nature and Its Development* (1777) is typically singled out as the likely resource for Kant's subsequent discussion of the imagination.[249] Tetens, alongside Lambert, has almost invariably been characterized as an eclectic, a thinker falling out of the usual sets of categories and allegiances. Like Lambert, Tetens was interested in mediating between the positions that had been established by Locke and Leibniz—mediating, that is, between an acceptance of Locke's position regarding the matter of cognition and the need, with Leibniz, to discover the universal basis of conceptual form.[250] But whereas the mathematical sciences of astronomy and optics had been influential for Lambert's philosophical development, Tetens would turn to the life sciences, and theories of generation in particular, when searching for "analogies" by which to understand the nature of the mind. In the 1770s Tetens was actively engaged in philosophical work and he accepted a professorship in philosophy at Kiel University in 1776. By the 1780s, however, Tetens's interests had shifted to actuarial practices, with the publication of a major two-volume work on the topic in 1785–1786, securing his place as a forerunner in the history of that science.[251] After 1789 Tetens left academia altogether, moving to Copenhagen and finishing his career as a high government finance officer in charge of the Danish insurance plans he had created for retirees and widows.

Tetens's move away from philosophy would prove to be especially disappointing to Kant, who included Tetens alongside Herz, Mendelssohn, and Garve as those upon whom he had "counted most" (10:270) for promoting his work.[252] It was reasonable for Kant to have expected that Tetens would take an interest in the first *Critique*, for in 1775 Tetens had cited Kant's *Inaugural Dissertation* several times in his own work *On General Speculative Philosophy*, a book that was oriented by many of the same themes that had been addressed by Kant in his *Dissertation* and was explicitly devoted to the question of the possibility of establishing metaphysics as a science at all.[253] When Tetens's *Philosophical Essays* appeared in 1777, Kant seems to have gone through the work carefully, above all the second volume, where in addition to the nature of the soul, Tetens discussed "Spontaneity and Freedom," devoting some five hundred pages to "Man's Perfectibility and Development" so far as these could be understood on the biological model of what Tetens described as an "evolution through epigenesis" (*Evolution durch Epigenesis*)—a position said by Tetens to have been developed by him primarily under the influence of Bonnet.[254]

As Tetens saw it, his borrowing from Bonnet amounted to a kind of restoration of the latter's real position, given that he took the new generation of psychologists to have distorted Bonnet's views in their effort to reduce psychic events to the material workings of oscillating fibers, vibrating nerves, and every other manner by which "material ideas" could be traced.[255] This kind of material reductionism was in the end as speculative as traditional metaphysical investigations into the properties of the soul had been. According to Tetens, there were two reasons for reaching this conclusion: first, mechanism claimed a descriptive capacity where it in fact had none and, second, it was always forced in the end to return to some kind of appeal to the immaterial soul at the seat of cognition. For these psychologists-cum-metaphysicians, as he put it, "the brain [*Denkorgan*] is a machine and the soul is the source of its power."[256] Tetens urged his readers to use organic analogies instead of mechanical ones when investigating cognition and to recognize that theories of generation were of special value for this investigation. Thus Tetens asked, "Can the formation [*Ausbildung*] and development [*Entwickelung*] of the soul, the emergence [*Entstehung*] of a sequence of ideas, and the growth [*Wachsen*] of the entire inner system of thought—the origin of completeness [*Ursprung der Fertigkeiten*] and so on and so forth—insofar as these are all based on the corporeal brain, [can these] not be represented as occurring in a similar manner as the formation, development, and the growth of the organic body?"[257] There were similar processes of organic formation occurring in the body, the brain, and the soul, Tetens explained, but the brain, as the "organ of the soul," preceded the soul's development, existing like an embryo in an enfolded (*eingewickelten*) state with only the potential, the mere "disposition" (*Anlagen*) to develop a soul and begin collecting impressions and ideas. Tetens considered the emergence of the soul from its previous state as mere potentiality to be the result of active generation even if its formation was originally directed by the brain. It was this attempt to balance Bonnet's preexistence theory (as the source of form) against the *vis essentialis* of Caspar Wolff's account (as the source of active force) that led Tetens to describe the organic generation of the soul as a case of *"Evolution durch Epigenesis."*

The importance of Kant's interest in Tetens during these years cannot be overestimated. And Hamann's oft-quoted remark to Herder in 1779 regarding Kant's attention to Tetens's work is thus worth repeating again—"Kant is hard at work on his Moral [*sic*] of Pure Reason and Tetens lies open constantly before him"—but it is in fact Kant's specific complaints regarding Tetens during this period that need to be held foremost in mind.[258] As Kant wrote about the matter to Marcus Herz at the time,

> Tetens, in his diffuse work on human nature, made some penetrating points; but it certainly looks as if for the most part he let his work be published just as he wrote it down, without corrections. When he wrote his long essay on freedom in the second volume, he must have kept hoping that he would find his way out of this labyrinth by means of certain ideas that he had hastily sketched for himself, or so it seems to me. After exhausting himself and his reader, he left the matter just as he had found it, advising his reader to consult his own feelings. (10:232)

Kant's private reflections during this period are equally significant for the purposes of discerning his reaction to Tetens's theory of cognition. "I am not concerned with the evolution [*Evolution*] of concepts like Tetens," Kant declared, "nor with their analysis like Lambert," indeed "I stand in no competition with these men." The difference, something

Kant took to be placing him in a different contest altogether, was his own emphasis on the objective validity of the intellectual concepts: "Tetens investigates the concepts of pure reason merely subjectively (human nature), I objectively. The one analysis is empirical, the other transcendental" (18:23).

Kant's appeal to the objective validity of the intellectual concepts, and the identification of his own method as specifically "transcendental," sound reassuringly familiar when looked at from the perspective of the *Critique of Pure Reason*. It is, however, only by paying attention to the precise nature of Kant's distinction between himself and Tetens at this juncture that we can make sense of the work left to be done by Kant before his work on the *Critique of Pure Reason* could be completed. For by charging Tetens with an account that was empirical so far as it was based upon human nature, Kant seems to have discovered a gap in his own account thus far. To be precise, Tetens had turned to the life sciences for his models when explaining the genesis of concepts and indeed the mind itself; reading Tetens, Kant realized that distinguishing his theory of cognition from the approach adopted in the *Philosophical Essays* would require something more than relying on the epigenetic origin of cognition. In the same manner that the nature of the originally acquired intellectual concepts had been reconceived since 1770—most importantly, insofar as they had become identified since then with the rules for judgment formation—the weight placed by Kant on their epigenetic origin would now have to be reconceived to the extent that the special account of their origin could be balanced against a description of their transcendental functions. Kant knew that his own "dissection of the faculty of understanding" (A66/B90) would move far beyond the theories provided by Locke and the empirical psychologists like Condillac and Bonnet who had followed him. But the similarities between Tetens's discussion and the still-developing transcendental account forced Kant to recognize that his own position could be mistaken for a physiological investigation into the origin of cognition. Indeed, because physiology fell under the domain of empirical psychology—as per the taxonomy of metaphysics—the entire project threatened to finally reduce itself to an empirical investigation, and this would certainly not provide Kant with the means by which he could rescue metaphysics from the threat of subreption.

Developing an adequate response to Tetens's work would prove to be extremely challenging for Kant, but the metaphysics lectures he gave during this period seem to have provided him with the platform

CHAPTER SIX

he needed in order to stake out a preliminary position. The need to distance himself from Tetens's approach to the contents of cognition would drive Kant to draw a distinction between the cognitive acts whose contents were sensible and those acts whose contents were the pure a priori components of cognition. In the lectures Kant would distinguish these in terms of a difference between "empirical" and "rational" cognition; in the first *Critique* this distinction would appear as a difference between the contents of "reproductive imagination" and the a priori contents of the "transcendental productive imagination" that served to ground empirical cognition. In this way Kant hoped to emphasize the transcendental framework within which the new theory of cognition was supposed to be operating in its production of objectively valid knowledge. As for the special account of the categories' epigenetic origin, indeed as for the centrality of Kant's attention to the question of origin throughout his account, this simply could not be eliminated. It would be in fact Kant's efforts to deemphasize it that would lead to much of the confusion and many of the complaints regarding Kant's obscurity in the "deduction" of the categories at the heart of his account. As Kant himself knew, the problem with the physiological approach was not its attention to origin per se; it was the genealogy proposed by the empiricists that was problematic insofar as they traced the lineage of all knowledge claims back to experience (Aix). The real genealogy of knowledge, as readers of Kant's *Critique of Pure Reason* would soon discover, led to its origin within reason itself. For this to make sense, however, we must first return to Kant at the end of the 1770s as he struggled to develop a response to Tetens and the *Philosophical Essays*.

From Empirical Psychology to a Transcendental Theory of Imagination

The discussion of empirical psychology had long been a central component in Kant's lectures on metaphysics. Following the textbook for the course, Baumgarten's *Metaphysica*, Kant divided the semester between considerations of "ontology," "cosmology," "empirical psychology," "rational psychology," and "theology."[259] Among these, empirical psychology covered the most ground in the lectures. As Kant defined its scope, it became immediately clear that physiology and psychology were to be identified insofar as each represented a form of cognition that took experience as its starting point: "Empirical physiology is the cognition of objects insofar as it [cognition] is obtained from principles of experience" (28:221), Kant explained, and "empirical psychology is the cognition of the objects of inner sense insofar as it [cognition] is obtained from experience" (28:222). Given the origin of its objects, empirical psychology belonged no more to metaphysics than empirical physics did. Kant took empirical psychology's long-standing presence within metaphysics to be easily explainable, however, since in the past there had been no precise boundaries for determining what should or should not be included in a metaphysical investigation. If only empirical psychology could be developed into an independent area of study, Kant suggested, then its investigations would contain as much potential as those of empirical physics, for the contents of empirical psychology constituted significant portions of anthropology, and this aspect of human life could then be investigated in the same manner as was being done in the natural histories of plants and animals (28:224). As Kant would put this all a few years later, "Though it [empirical psychology] is but a stranger [to metaphysics] it has long been accepted as a member of the household, and we allow it to stay for some time longer, until it is in a position to set up an establishment of its own in a complete anthropology, the pendant to the empirical doctrine of nature" (A849/B877).[260]

Given the size of its domain, the portion of Kant's lectures devoted to empirical psychology treated a large number of topics, and these generally followed, at least in outline, the traditional path taken by metaphysicians with respect to the soul. Thus there was an initial discussion of the nature of the soul as substantial, simple, immaterial, and intelligent, followed by consideration of the soul's activity in forming representations, of its aesthetic feelings of pleasure and displeasure, of its will or "faculty of desire," and finally, of its connection to the body. Because Kant traced the contents of the latter considerations to sense experience, he took them to be distinguishable from the proper concerns of metaphysics.[261]

Kant's lectures on metaphysics would be changed, however, once Tetens's *Philosophical Essays* were published.[262] For his next set of lectures Kant would add a fresh set of distinctions. In contrast to empirical physiology, for example, *rational physiology* was introduced as "the cognition of objects insofar as it is obtained not from experience, but rather from a concept of reason" (28:221), and in contrast to empirical psychology, *rational psychology* was introduced as an area belonging to metaphysics absolutely, since its principles were "borrowed from pure reason" (28:223). The basis for these distinctions would be repeated in

CHAPTER SIX

Kant's subsequent classification of all the faculties of cognition into their lower and higher forms. The lower faculties of representation, desire, pleasure, and displeasure were fundamentally passive, depending on experience for their contents. The higher faculties, by comparison, were described as "self-active": the higher faculty of cognition was "a power to have representations from ourselves," the higher faculty of desire was "a power to desire something from ourselves independently of objects" (28:228–229), and the higher faculty of pleasure and displeasure looked to the self as well, in this case toward its feelings of either the promotion or hindrance of life (28:247).

Despite the lower faculty of representation's "passivity" regarding the origin of it contents, however—and much like sensibility's synthesis of *apparentia* in the 1770 *Dissertation*—this faculty was still tasked with the formation of cognitions; indeed it was described in this capacity as a "formative power" (*bildende Kraft*), one arising from the "spontaneity of the mind" (28:230). The work of formation (*Bildung*) served as the root of Kant's discussion of this power. There was the power to illustrate the present (*Abbildungskraft*), to reproduce the past (*Nachbildungskraft*, also called the *Imagination*), and to anticipate the future (*Vorbildungskraft*). In addition to these basic powers of formation within a temporal horizon, Kant described a faculty of imagination (*Einbildung*, also called sensible *Dichtungskraft*), which was capable of producing images that had not been borrowed from experience; a faculty of correlation (*Gegenbildung*) responsible for analogical or symbolic formation; and a faculty of cultivation (*Ausbildung*), which contained a drive to cultivate and complete everything.[263] "All of these acts [*actus*] of the formative power," Kant explained, "can happen voluntarily and also involuntarily. Insofar as they happen involuntarily they belong wholly to sensibility; but so far as they happen voluntarily, they belong to the higher faculty of cognition" (28:237). Despite their great activity, it was because the lower faculties of formation worked on material whose source was external to themselves, according to Kant, that they could not be described as the true "author of their representations" in the manner of the higher faculties.

With the division between the lower and higher faculties tied thus far to the different sources of content, it was clear Kant still remained within the framework established by the *Inaugural Dissertation* regarding the grounds for differentiating between the separate faculties of sense and intellect. The advance since 1770 concerned the manner by which Kant now sought to connect them. As the first step to this, remember that by 1773 Kant had identified the originally generated intellectual concepts with the rules for logical subordination and that he took the connection between concepts and objects to be thereby established insofar as the rules for judgment formation operated at the same time as the means for generating representations of objects. As Kant had put it then, "The concepts of understanding express all the *actus* of the powers of the mind, insofar as representations are possible in accordance with their universal laws, and indeed their possibility *a priori*" (17:622). But if this had solved the problem of the so-called objective deduction, it left the "subjective" deduction, or the account of the means by which this was all supposed to take place, unaddressed. Kant's attempt to address this and thereby complete the account emerged now in the role played by the formative power as the means for connecting the activities of sense and intellect or, as they appeared here in the metaphysics lectures, of the lower and higher faculties.

Unfortunately the records of Kant's attempt at an explanation demonstrate the degree to which he still remained uncertain in these lectures of the precise means for connecting the various operations being performed by the cognitive faculties. The main elements of Kant's eventual account had, however, been assembled—space and time, the understanding, reason, spontaneity, judgment—only the "transcendental imagination" was missing, replaced in this case by the "voluntary" agency of the formative power. As Kant saw it, the formative power operated at both levels of cognition: "We have cognitions of objects of intuition by virtue of the formative power, which is between the understanding and sensibility. If this formative power is in the abstract, then it is the understanding. The conditions and actions of the formative power, taken in the abstract, are pure concepts of the understanding" (28:239). In this manner conceptual determination was linked—as a process of formation—to the active work being done in the formation of representations at the empirical level. But Kant took the "abstract" (*in abstracto*) nature of conceptual determination to be critical, given that he continued to maintain the isomorphism of the intellectual concepts and the rules for judgment formation. In his words,

The higher faculty of cognition is also called the understanding, in the general sense. In this meaning the understanding is the faculty of concepts, or also the faculty of judgments, but also the faculty of rules. All three of these definitions are the same, for a concept is a cognition which can serve as a predicate in a possible judgment. But a judgment is a representation of the comparison with a general feature, and a concept is a general feature. But a judgment is also always a rule, for a rule gives the relation of the particular to the general. (28:240)

CHAPTER SIX

Kant's insistence on an essential identity underlying the various parts of cognition would reappear in a similar vein, for example, in his subsequent description of the unity of apperception: "We have already defined the understanding in various different ways: as a spontaneity of knowledge (in distinction from the receptivity of sensibility), as a power of thought, as a faculty of concepts, or again of judgments. All these definitions, when they are adequately understood, are identical. We may now [also] characterize it as a *faculty of rules*" (A126). Kant's attempt, moreover, to connect the lower and higher faculties by means of the synthetic work being done by the formative power, and his identification, *"in abstracto,"* of the understanding and the higher use of the formative power would also reappear, as when he would later announce, "The unity of the apperception in relation to the synthesis of imagination is the understanding; and this same unity, with reference to the transcendental synthesis of imagination, the pure understanding" (A119).[264]

Tetens had also distinguished between lower and higher levels of formation when discussing representations in the *Philosophical Essays*. He too had drawn distinctions between the mere reproduction of images by the imagination (*die Einbildungskraft oder Phantasie*) according to the "psychological" laws of association and the "self-active," creative production of representations via the formative power (*die bildende Dichtkraft* or *Dichtungsvermögen*). But throughout it all, Tetens had also maintained a general fidelity to Locke's empiricism regarding experience as the singular source of content.[265] Kant wanted something different: he wanted to shift the priority placed by the empiricists on experience to cognition instead as providing the only objective grounds for experience. As Kant had remarked, his investigation would be transcendental, not empirical (18:23), and his discussion of cognition would therefore need to emphasize its transcendental role in the production of knowledge.

By 1780, Kant was ready to reprise the account offered in the lectures on metaphysics in terms that would sound close to the final version offered in the *Critique of Pure Reason* one year later. The so-called lower faculty of formation was now explicitly identified as the reproductive imagination (*Einbildungskraft*) so far as it rested on the synthesis of intuitions that had been empirically apprehended. It became productive when generating analogies based on a previous "synthesis of apprehension," and it became pure so far as it was connected to an object generated by space and time as the forms of pure sensible intuition. But "the transcendental synthesis of imagination" itself rested on (*geht bloß auf*) "the unity of apperception in the synthesis of the manifold in general" (23:18). The various grades of the imagination—reproductive, productive, pure, transcendental—became thereby so many steps between the sensible content of intuition and the unity of apperception as the highest ground for the possibility of experience. Against the empiricists, therefore, appearances were meaningful not because they were first met with in the senses, but only insofar as they could be connected to apperception, a connection that depended upon their first being conceptually determined by the transcendental imagination as representations belonging to a unified consciousness (23:19).[266] This shift from the empirical to the transcendental—a shift culminating in the expulsion of the reproductive imagination from the transcendental deduction altogether in 1787—demonstrated Kant's efforts to combat the identification of Tetens's account and his own so far as Tetens's empiricism was concerned. This would not, however, be the end of Kant's response to the *Philosophical Essays*. Tetens had undertaken his investigation into the origin of knowledge by way of models supplied by the life sciences, and in this sense Tetens's work contained significant points of contact with Kant's own theory of cognition. Kant would thus have to distinguish his own project from Tetens's physiological approach. The only question was how.

Transcendental Philosophy and the Physiology of Pure Reason

For all the discussions of Buffon, Bonnet, Wolff, and the other generation theorists that Tetens had considered in the *Philosophical Essays*, Kant still thought that the best way to view Tetens's position was through the lens provided by Locke's theory of cognition. Tetens followed Locke's physiological approach, according to Kant, since Tetens turned to the laws of association when explaining specific mental processes. This was true both for Tetens's manner of treating objects of experience and for his attempt to understand the processes of cognition itself. It was the effort to distinguish himself from this kind of empiricism that had led Kant in his metaphysics lectures to draw a distinction between empirical and rational psychology on the basis of their respective contents: principles of experience made up empirical psychology, and concepts of reason were the objects of rational psychology. By the 1780s, however, Kant would restructure his discussion of empirical psychology altogether when delivering his lectures. Discussion of the will in terms of its empirical motives and inclinations, for example, would henceforth be defined as belonging to "pragmatic anthropology," a discipline oriented by the same sets of practical rules and prudential maxims that were central to the work being done by the moral empiricists. Those portions of Kant's lectures devoted to the characteristics of the soul as simple, substantial, connected to the body, and so on would now appear as cases of false inferences in line with Kant's discussion of them in the *Critique of Pure Reason*. As Kant described the illusion generated by this kind of inference in the sections devoted to this in the *Critique*, it stemmed from the mistake of asserting reason's ideas regarding the nature of the soul—ideas stemming from the same concepts of reason that had formed the content of rational psychology in the earlier lectures—to be in fact positive descriptions of the soul itself, descriptions that were then used by reason to form the positive basis of rational psychology as a "doctrine." The only discussion of the soul that Kant was willing to grant was a description of what he variously referred to as the *cogito* or the "bare I think," an indeterminate sense of oneself as a thinking thing.[267] And the most that introspection could yield, according to Kant, was an empirical inventory of the contents of our thoughts:

> If our knowledge of thinking beings in general, by means of pure reason, were based on more than the *cogito*, if we likewise made use of observations concerning the play of our thoughts and the natural laws of the thinking self to be derived from these thoughts, there would arise an empirical psychology, which would be a kind of *physiology* of inner sense, capable perhaps of explaining the appearances of inner sense, but never of revealing such properties as do not in any way belong to possible experience (e.g., the properties of the simple), nor of yielding any *apodeictic* knowledge regarding the nature of thinking beings in general. It would not, therefore, be a *rational* psychology. (A347/B405)

Empirical psychology as "a kind of physiology of the inner sense" was thus far less detrimental than rational psychology, whose false doctrines regarding the soul offered a perfect example of the kind of subreptive logic that had been undermining metaphysics from the start. There was, however, a negative function that could be provided by a chastened rational psychology once it was reoriented as a "discipline" meant to curb presumptions regarding the soul: "It keeps us, on the one hand, from throwing ourselves into the arms of a soulless materialism," Kant explained, and "on the other hand, from losing ourselves in a spiritualism which must be quite unfounded so long as we remain in this present life" (B421). Once rational psychology was prevented from the subreptive transcendental employment of its ideas, in other words, it could be made valuable by its practical employment within the moral sphere. In this sphere, ideas of the soul could "regulate our actions as if our destiny reached infinitely far beyond experience, therefore far beyond this present life" (B421, cf. 5:461). Empirical psychology, by contrast, would only become valuable as the focus of a practical anthropology that remained to be established.

Reviewing in order the topics that had earlier fallen under the heading of empirical psychology in the lectures on metaphysics, one can see that in the 1780s Kant would divide these such that consideration of the properties of the soul and its connection to the body would be taken up in the account of the paralogisms in the *Critique*; that discussion of the involuntary and voluntary use of the will would be reassigned to the anthropology lectures and the moral works; and finally, that the discussion of aesthetics would subsequently become a discussion of taste and beauty in either an anthropological context or as part of a critique of aesthetic judgment. The remaining subject of empirical psychology was a consideration of the soul's activity in forming representations. And this account was linked for Kant, as just shown again above, to the empirical psychologist's physiological approach to the processes of cognition.

In 1781, empirical physiology continued to refer for Kant to the uncritical appeal to experience as the genealogical origin of knowledge. In the opening pages of the *Critique of Pure Reason* Kant characterized this approach in terms of the initial hope that had been offered investigators of both nature and knowledge by Locke's *Essay*. It was, however, hope raised in vain, for by looking to experience as a ground, Locke had gotten the genealogy all wrong. As Kant reprised this history,

> In more recent times, it has seemed as if an end might be put to all these controversies and the claims of metaphysics receive final judgment, through a certain *physiology* of the human understanding—that of the celebrated Locke. But it has turned out to be quite otherwise. For however the attempt be made to place doubt upon the pretensions of the supposed Queen by tracing her lineage to vulgar origins in common experience, this genealogy has, as a matter of fact, been fictitiously invented, and she has continued to uphold her claims. (Aix)

Kant would go on to contrast empirical physiology, as an unreflective appeal to experience as given, to a rational physiology that had been newly defined as the critical appeal to experience as only the result of its prior determination according to the transcendental conditions

CHAPTER SIX

for its possibility. This kind of "immanent physiology," as Kant also referred to it, "views nature as the sum of all objects of the senses, and therefore just as it is given to us, but solely in accordance with *a priori* conditions, under which alone it can ever be given us" (A846/B874).

In the *Critique of Pure Reason*, empirical physiology still served as a counterpart to empirical psychology but now it was because neither recognized the basic tenets of transcendental idealism as having set the grounds for both the possibility of the experience of nature and the cognitive processes by which nature was empirically apprehended at all. The empirical physiologist traced ideas back to nature, but to a nature naively taken to be knowable apart from any epistemic conditions set by the knower. Empirical psychology, as an inventory of the contents of inner sense, similarly pointed outside of itself for its contents. The empirical psychologist, as Kant would have put it, could be said to have relied upon the reproductive imagination without recognizing the higher functioning of the transcendental imagination as its a priori basis.

In Kant's 1782 lectures on metaphysics he again addressed the difference between empirical and critical approaches to the problem, referring once more to Locke. "Physiology of pure reason is the inquiry into the origin of concepts," Kant explained. "It is the consideration of the nature of reason: how reason generates [*erzeugt*] concepts of the understanding." As such, it is "an investigation of a matter of fact, it is, as the lawyers say, a question of fact [*quaestio facti*]." Such an investigation might be subtle, Kant argued, but it did not belong in metaphysics. Metaphysics started from the fact that concepts were acquired (*acquisiti*) through the pure use of reason, but then it asked, more importantly, by what right it could use them (29:764). This was the question of right (*quaestio iuris*), and for Kant it went beyond the mere physiology of pure reason to its critique. Following this remark, Kant went on to recount the history of physiological approaches for his students:

The former question [of fact] has been the business of two philosophers, of Locke and Leibniz, the former wrote a book on human understanding, *de intellecto humano*, the latter published a book with this title in French. Locke adheres to Aristotle and maintains that concepts arose from experience through acts of reflection. Leibniz adheres to Plato, but not to his mysticism, and says that the concepts of the understanding are prior to any acquaintance with the sensible understanding. . . . No one has thought of a critique of pure reason until now. (29:764)

With its focus on the origin of concepts, Kant concluded, physiology "is really a part of psychology," and it had to be distinguished sharply from critique (29:764). While the opening passages of the *Critique* had focused on the empiricist's mistaken genealogy regarding the *source* of ideas, in 1782 Kant seemed to have broadened his critique of physiology to include all inquiries concerned with an investigation into the origin of concepts.

But in this instance Kant had clearly overstated the case against physiology. For Kant too was deeply concerned with the "question of fact" regarding the origin of concepts; indeed their epigenetic generation had been a central component of his developing theory of cognition since 1770. Kant needed something to distinguish his account from that of the physiologists—by this definition, Locke, Tetens, even Leibniz—besides an attention to the question of origin, and it was for this reason that he had worked in the deduction of the categories of experience to balance the importance of the question of their origin with their transcendental capacity to provide objectively valid knowledge. As for the specter of physiology, Kant's solution had been to rehabilitate a redefined "rational physiology"—while still criticizing Locke and others as physiologists—as a respectable alternative to empirical physiology given rational physiology's attention to the transcendental grounds of experience. It was in this sense that Kant could say, "Metaphysics, in the narrower meaning of the term, consists of *transcendental philosophy* and *physiology* of pure reason" (A845/B873). "Transcendental philosophy," in other words, regarded reason and understanding as operating in a system of concepts and principles connected only to objects in general, and rational physiology, as a discipline, regarded nature in accordance with the a priori conditions established by transcendental philosophy; it was because of this that physiology was able to cover both a "rational physics" of corporeal nature and a "rational psychology" for the soul (A846/B874). How could rational physiology claim a priori knowledge of nature when it depended upon objects given in an a posteriori manner? "The answer," Kant replied, is that "we take nothing more from experience than is required to *give* us an object of outer or of inner sense. The object of outer sense we obtain through the mere concept of matter (impenetrable, lifeless extension), the object of inner sense through the concept of a thinking being (in the empirical representation, 'I think')" (A848/B876). In this formulation physiology was still linked to experience, but it belonged to metaphysics nonetheless, according to Kant, because it had been grounded by the work of transcendental philosophy. Thus although Kant continued to be critical of physiology for its connection to empirical psychology in the *Critique of Pure Reason* (e.g., A347/B405, Aix), he made room for rational

CHAPTER SIX

physiology as a transcendentally grounded account of nature, and he reoriented rational psychology as the only safe route toward the practical employment of ideas regarding the soul.

Kant's proposed distinction between questions of fact regarding the origin of knowledge, and questions of right regarding the justified use of epistemic criteria when generating knowledge, as the key for distinguishing physiology from critical philosophy in the 1782 lectures was thus inconsistent with the work done in the *Critique of Pure Reason*—indeed, once Kant turned to the discussion of "Ontology" in the same 1782 lecture course, the familiar emphasis on the need to locate the origin of the intellectual concepts returned, for whereas "logic deals with the connections of concepts," Kant explained, "metaphysics [deals] with their origin" (29:802)—but it was also the remnant of Kant's least successful attempt to displace the question of origin from his account. Critical philosophy was uniquely indebted to an account of the epigenetic generation of concepts, which is why Kant's answer to the question of fact or origin in the deduction would ineluctably serve as the real ground for his proof with respect to the question of the categories' rightful application to experience.[268] An account of this is the focus of the next chapter, but before we leave our discussion of Tetens, it is important to see that by reframing the deduction as a genealogical proof Kant did not by any means intend to naturalize his account of reason. Kant would take the epigenesis of reason to be real but only in a *metaphysical* sense, and this in the end was what finally distinguished him from Tetens. Tetens was an empirical psychologist, and in his physiological investigation of cognition he not only turned to biological analogs when discussing the generation of ideas, he deliberately sought thereby to present a naturalized psychology. Kant, by contrast, was in the end a metaphysician, and his own species of organicism would therefore have to be nonnaturalistic when it came to reason and the processes of cognition.

SEVEN

Kant's Architectonic: System and Organism in the *Critique of Pure Reason*

The Doctrine of Method: The *Bauplan* of the System

Kant's transcendental deduction has been dogged from the start as a piece of excessive obscurity. As one early critic summarized the problem, the obscurity of Kant's work was at its greatest in the transcendental deduction, even as it was "this part of the *Critique* that should be the clearest, if the Kantian system is to afford complete conviction."[269] When responding to this charge, however, Kant maintained his support for the deduction, explaining its difficulty as a defect owing only to "the manner of presentation and not the ground of explanation" (4:476). But what was Kant's actual ground of explanation in the transcendental deduction? It is my contention that in order to uncover this ground one needs to begin with the final sections of the *Critique of Pure Reason* if there is to be any hope of eliminating the obscurities associated with the deduction. For what one sees in these later discussions, especially in the case of the architectonic, is the overriding importance of organic models for Kant's conception of reason, an organicism that must remain in focus if the deduction is to be brought out from the depths of its supposed obscurity and the actual ground of its explanation revealed for what it is.

Beginning at the end of Kant's book, then, we discover

that in the closing pages of the *Critique* Kant decided to walk his readers through the "History of Reason," a landscape littered with structures that had turned into ruins. During the course of this history there had been three issues, according to Kant, that were at stake during each of the chief revolutions in metaphysics. The first issue concerned the *object* of knowledge; the second, the *origin* of the modes of knowledge; and the third, the *method* of metaphysics.[270] Taking up these issues in order, Kant presented his usual cast of characters, now divided between "intellectualists" (Plato) and "sensualists" (Epicurus) with reference to the *object* of knowledge, between "noologists" (Plato and Leibniz) and "empiricists" (Aristotle and Locke) with respect to the *origin* of knowledge, and finally between "dogmatists" (Wolff) and "skeptics" (Hume) in terms of the *method* of investigation. The first two issues were of course intimately connected, and Kant's own path toward understanding them had depended in part on a synthesis of the positions put forward by the opposing camps. The object of knowledge had to be understood as the synthetic result of the intellectual and the sensual together, for, according to Kant, only this kind of transcendental imposition of form on matter could yield knowledge whose certainty was guaranteed.

As for the third issue revolutionizing metaphysics, the issue of method, Kant had already made it clear that critical philosophy offered the only method capable of providing a suitable "dwelling place," as he put it, for metaphysics, a site capable of providing safety and shelter in the face of "marauding nomads" but whose modest offerings meant that it would be embraced only after reason had tried and failed to set up a home elsewhere. In describing this Kant likened the process of reason's failed attempts to secure shelter to a natural process of growth and maturation—that is, to a movement through the stages of reason's infancy (dogmatism), youth (skepticism), and maturity (criticism). Thus he characterized reason's development as follows:

The first step in matters of pure reason, marking its infancy, is *dogmatic*. The second step is *skeptical*, and indicates that experience has rendered our judgment wiser and more circumspect. But a third step, such as can be taken only by fully matured judgment, based on assured principles of proved universality, is now necessary.... Skepticism is thus a resting place for human reason, where it can reflect upon its dogmatic wanderings and make survey of the region in which it finds itself.... But it is no dwelling place for permanent settlement. Such can be obtained only through perfect certainty in our knowledge, alike of the objects themselves and of the limits within which all our knowledge of objects is enclosed. (A761/B789)

The materials making up a permanent settlement for a matured reason, as Kant saw it, were the same items that had led to so much confusion in the history of reason itself: sensible intuition, intellectual concepts, and the ideas of reason. Only a proper methodological approach could hope to safely construct a home for metaphysics from materials that had led previous architects of metaphysics astray. "Although we had contemplated building a tower which should reach to the heavens," Kant explained while surveying the results of his labor, "the supply of materials suffices only for a dwelling house, just sufficiently commodious for our business on the level of experience, and just sufficiently high to allow of our overlooking it" (A707/B735).

Kant's discussion of the building materials in use within the systems of metaphysics marked a transition in the *Critique* from the "Doctrine of Elements" to the "Doctrine of Method."[271] It was a transition that meant moving from a material consideration of the elements and faculties described in the major sections devoted to intuition, concepts, and ideas to an account of the system of reason itself. As Kant turned to a consideration of the system of reason, his vocabulary underwent a change, shifting from the language of construction and building materials to descriptions borrowed from the language of organic growth and development. It was a linguistic transition that had been introduced by way of the biography of reason's development from infancy to adulthood, an organic course of formation that, as Kant now explained it, had been a case of the "sheer self-development of reason." Rehearsing the course of reason's maturation, Kant explained,

Systems seem to be formed in the manner of lowly organisms, through a *generatio aequivoca* from the mere confluence of assembled concepts, at first imperfect, and only gradually attaining to completeness, although they have one and all had their schema, as the original germ, in the sheer self-development of reason. Hence, not only is each system articulated in accordance with an idea, but they are one and all organically united in a system of human knowledge, as members of one whole, and so as admitting of an architectonic of all human knowledge. (A835/B863)

What the history of reason demonstrated for Kant was that all attempts at metaphysics had been "organically united," that they were connected by virtue of their common origin in the germ of reason, and that they had been differentiated only as part of reason's own path of self-development. The history of reason thus provided its investigators with a genuine natural history, for each of its varieties could be traced in their entirety to their point of origin, a common descent that had been easy to overlook given the enormous modifications taking place in the history of the species as a whole. As varieties of reason, the systems of metaphysics functioned organically, like "members of one whole," so Kant could be precise when describing the manner by which reason had grown into a unified system. As he defined this organic growth, "The whole is thus an organized unity (*articulatio*), and not an aggregate (*coacervatio*). It may grow from within (*per intussusceptionem*), but not by external addition (*per appositionem*). It is thus like an animal body, the growth of which is not by the addition of a new member, but by the rendering of each member, without change of proportion, stronger and more effective for its purposes" (A833/B862).[272] Kant believed that the connection between the parts of the system could be likened to the organic interworking of the organs in an animal body because the unity of the system, like the unity of an organism, determined not only the exact number and placement of its members but the end toward which they aimed. In each of these cases this was an end that had been reflexively defined from the start; in the case of reason it had been contained within the system as an idea of its completion from the very first moment of its self-conception. The end of the history of reason, that is, its idea of itself as a fully developed whole, was originally present within reason—present as an "original germ in the sheer self-development of reason"—a germ or idea that both set the goal for reason's completion and somehow also grounded the possibility of its actual achievement.[273]

If the history of reason presented a whole, that was because its member systems could trace their ancestry back to an idea of reason's self-completion. It was a whole, in Kant's words, "in view of the affinity of its parts and of their derivation from a single supreme and inner end, through which the whole is first made possible" (A834/B862). By appealing to the development of reason in this manner, the teleological course taken during the natural history of reason took on the cast of destiny, such that Kant could open his *Critique of Pure Reason* with the announcement that "human reason has this peculiar fate that in one species [*Gattung*] of its knowledge it is burdened by questions which, as prescribed by the very nature of reason itself, it is not able to ignore, but which, as transcending all its powers, it is also not able to answer" (Avii). It was in fact because of its peculiar fate that the history of reason had been methodologically dominated by either dogmatic prescriptions or skeptical dismissals with respect to these questions, but it was nonetheless reason's unavoidable destiny both to pose these questions and to proceed through identifiable stages of development—

its dogmatic infancy, its skeptical youth—all as part of its drive toward completion.

Kant was thus clear when it came to locating the value of the *Critique of Pure Reason* with respect to the history of reason, for the critical approach to metaphysics offered a negative yet necessary service in support of reason's continual advance toward its completion (A710/B738; cf. 4:368). If this negative service of disciplining reason could be combined with reason's practical employment, according to Kant, then critical metaphysics would itself become "the full and complete development [*Kultur*] of human reason" (A850/B878).[274] As Kant put the point a couple of years later in the *Prolegomena*, "Such is the end and use of this natural predisposition [*Naturanlage*] of our reason, which gave birth to metaphysics [*ausgeboren hat*] as her favorite child, and whose generation [*Erzeugung*], like every other in the world, is not to be ascribed to blind chance but to an original germ [*Keim*], wisely organized for great ends" (4:353).[275] The methodological difference between critical philosophy and the other philosophical systems was that critique began its investigations by tracing the genealogy of the questions themselves, questions that one and all led back to reason as their source.

Thus throughout the *Critique of Pure Reason* Kant would repeatedly remind his readers that the source of metaphysics could be traced to something rooted in the very nature of reason itself.[276] And indeed as Kant developed the point, the rootedness of these questions in reason became almost synonymous with the image of reason Kant drew. Reason could be best viewed as a root with two stems, branches that, when taken alone, had served as the platforms on which the various systems of reason had been erected. "By way of introduction," Kant could say at the outset of his investigation in the *Critique*, "we need only say that there are two stems [*Stämme*] of human knowledge, namely, *sensibility* and *understanding*, which perhaps spring from a common, but to us unknown root [*Wurzel*]" (A15/B29). By the end of his investigation, however, Kant was ready to identify the "unknown root," describing the work of the *Critique* in retrospect as having been the attempt to outline "all knowledge arising from *pure reason*," beginning from the point at which this "common root of our faculty of knowledge divides and throws out two stems, one of which is *reason*" as the whole higher faculty of knowledge, the other stem being "the empirical" (A835/B863).

What we can see here is that in attempting to capture the systematic unity of reason in its historical self-development, Kant was repeatedly drawn to organic imagery. Kant likened the system of reason to the organic unity of an animal, he took reason's historical development to be a movement from its infancy to its adulthood, and he described reason's function within this history as akin to that of a root. For reason was the root supporting its own historical development, a root whose stems were identical in their dependence upon reason for supplying the main source of their drive to know, even as the answers these systems reached would ultimately cause them to differentiate and even diverge, appearing in the end as varieties of dogmatism and skepticism throughout history. Within this history of reason, only critical philosophy had recognized that the organic affinity of sensibility and understanding within cognition mirrored the affinity of systems within the historical development of reason itself, that indeed this inner cognitive affinity was originary insofar as it was the unity of apperception that was reflected outward onto the history of reason itself.

Kant seems to have understood that botanical models were particularly suited to capturing this conception of reason, given that vegetative cycles of growth and propagation demonstrated precisely the fact of a whole within each of their parts and that they could capture by way of analogy, therefore, the simultaneity of unity and difference within the system of reason.[277] Kant found the image of the root to be perfect, moreover, not only for capturing reason's role as the basis from out of which everything else would grow but also for demonstrating the sense in which the questions posed by reason were impossible to remove given how deep they ran within the nature of reason itself. Reason was thus not only *disposed (als Naturanlage)* to ask metaphysical questions (B22), it was *compelled* by a need to ask them, a need so deeply rooted in its nature as to be inextirpable.[278] "The root of these disturbances," Kant exclaimed at one point, "which lies deep in the nature of human reason, must be removed. But how can we do so unless we give it freedom, nay, nourishment, to send out shoots so that it may discover itself to our eyes, and that it may then be entirely destroyed?" (A778/B806). Insofar as this root was intertwined with the fate of reason, Kant's own answer would show that complete destruction was not only impossible but ultimately unwanted from the perspective of reason's movement toward the achievement of its goals. What critical philosophy proposed instead, therefore, was containment of the problem. The negative work of the *Critique* could be achieved—and thereby any regression on reason's part to dogmatism and skepticism avoided—only by clipping the wings of a reason determined for lofty heights, returning it instead to the immanent use that was its proper domain. "Pure reason," as Kant stated the goals of criticism, should be "in fact occupied by nothing but itself" (A680/B708).[279]

The Transcendental Deduction: The *Bauplan* at Work

If Kant's conception of reason as the ground of unity—as ground of the unity of its history, of its faculties, and of experience itself—called to mind botanical images for him, the discussion of cognition was located in the vocabularies of origin and birth. It was therefore no accident that Kant deemed the centerpiece of his theory to be the transcendental deduction, for within the contemporary legal domain, a "deduction" was used primarily for determining questions of birthright and inheritance. As far as it would be positioned in the *Critique of Pure Reason*, the deduction was meant to trace the intellectual concepts to their point of origin in order to demonstrate their rightful application to experience. In terms borrowed from the court system, Kant thus explained, "Jurists, when speaking of rights and claims, distinguish in a legal action the question of right [*quid juris*] from the question of fact [*quid facti*]; and they demand that both be proved. Proof of the former, which has to state the right or the legal claim, they entitle the *deduction*" (A84/B117). In a legal deduction, research into genealogical lines—that is, discovering a certificate of birth in response to the question of fact—was the first step in proving the rights of a claimant. Distinguishing between competing claimants in order to determine rightful inheritance in response to the question of right was the second step of the investigation, and the summary presentation of the proof of rightful inheritance—a presentation that necessarily included the results of the first step's investigation into birthright—was referred to altogether as the legal "deduction" of a given claimant's rights.

But while this legal model seemed to set up the clearly defined stages of a relatively uncomplicated two-step procedure, Kant's actual proof of the rights to be granted the categories with respect to experience was less direct. Indeed, it would ultimately circle back upon itself insofar as a separate, deeper investigation into the birthplace of cognition as a whole was needed in order to declare reason the ultimate guarantor of any rights being later claimed by the categories themselves. In Kant's version of the deduction, therefore, the question of fact would ultimately be required to secure the question of right, and this was the case despite Kant's protestations regarding the relative independence and priority of the latter investigation.

The dominating significance of the *quid facti* for Kant's argument was not immediately obvious to Kant's readers as they attempted to follow the linear progression of the deduction itself. Part of this had to

CHAPTER SEVEN

do with Kant's decision to follow the legal model in distinguishing the two steps of his own investigation in terms of an initial inquiry into the origin of the categories, the so-called metaphysical deduction, and a subsequent proof of their necessity for experience in the transcendental deduction. But this division of labor between sections devoted to the origin of the categories on the one hand and their application to experience on the other was not the only problem for discerning the real nature of Kant's proof so far as its dependence upon the question of origin was concerned. A separate problem for interpretation emerged as a result of Kant's effort to distinguish a properly transcendental consideration of the conditions for the possibility of experience from the merely physiological investigation into the sources of knowledge, an investigation, in other words, that seemed to be especially tied to a discovery of origin and, on that basis, one that Kant was ready to dismiss. Thus on the heels of outlining the two-step procedure that his proof would undertake, Kant interrupted himself to announce the distance between his own transcendental account and the failed efforts on the part of physiologists to produce an empirical deduction. One might indeed feel "indebted to the celebrated Locke" for establishing a new line of investigation into the sources of knowledge, for example. "But," Kant argued, "a *deduction* of the pure *a priori* concepts can never be obtained in this manner; it is not to be looked for in any such direction. For in view of their subsequent employment, which has to be entirely independent of experience, they must be in a position to show a certificate of birth quite other than that of descent from experiences. Since this attempted physiological derivation concerns a *quaestio facti*, it cannot strictly be called a deduction; and I shall therefore entitle it the explanation of the *possession* of pure knowledge" (A86–87/B119). The problem facing Kant regarding this point, as was shown in the discussion of Kant and Tetens, was that Kant's investigation depended as much on the discovery of a "certificate of birth" as it had for the empiricists—in Kant's case, a certificate demonstrating the descent of the categories from neither experience nor God but indeed from reason alone. Kant's real argument against the empiricists' claims did not directly rest on a demonstration of the *quid juris*, that is, on a proof of the rightful claim to the objective necessity of the categories for experience at all. His approach in the *Critique* followed instead the same trajectory that his own thinking had taken since the *Inaugural Dissertation* in 1770, an intellectual trajectory, that is, that had been dominated by questions of origin and unity. Since this was the real path by which Kant would attempt to secure the possibility of necessity in experience, only direct attention to questions of origin and unity—and thus only indirect concern with the problem of the *quid juris*—would reveal the actual proof offered up by the deduction. As we will see, it would be a proof hinging upon an account of "transcendental affinity," the term Kant chose when referring to the organic unity of reason as the original ground of apperception.

When Kant introduced his discussion of affinity in the "Doctrine of Method," that is to say, in the section purported to be the *Bauplan* of the entire *Critique*, he opened with a reminder that within the history of reason, skepticism had represented an advance over dogmatism. This meant that Hume had started out "on the track of truth" and that it was important to discover, therefore, where this "most ingenious of all the skeptics" had become derailed (A764/B792). In Kant's diagnosis, Hume's rightly aimed skeptical attack on dogmatic metaphysics had stopped short of a solution in the form of a positive doctrine because Hume had failed to recognize the difference between "the well-grounded claims of the understanding and the dialectical pretensions of our reason" (A768/B796). By ignoring this distinction, Hume had reactively turned to experience as the sole source of our concepts and could not see, therefore, the difference between the empirical concepts operating in a posteriori judgments and the pure concepts whose origin lay in the understanding itself and who could thus serve as a basis for synthetic a priori judgments. As Kant summarized this, "Our skeptic did not distinguish these two kinds of judgments, as yet he ought to have done, but straightaway proceeded to treat this multiplication of concepts from out of itself and so to speak, the self-birth [*die Selbstgebärung*] of our understanding (and reason) without impregnation by experience [*ohne durch Erfahrung geschwängert*], as impossible" (A765/B793).[280] The unity of reason, an organic unity generated "without impregnation by experience," and from out of which all principles of affinity arose, served for Kant as the genealogical basis of an organic affinity between the parts of cognition. This organic affinity between the faculties meant there was a unity to apperception, a unity that served in turn as the transcendental condition for the possibility of a coherent experience. Kant could conclude on this basis, therefore, that *transcendental* affinity served as the condition for the possibility of an experience of *natural* affinity, that Hume, in other words, had been able to ascribe coherence to experience not as a result of the rules of association but rather only as a result of such natural affinity having been already produced by the transcendental affinity grounded by reason itself. In making this argument, Kant based the necessary connection of experience—the goal of investigations into the question of right—on a genealogical claim regarding reason as the common origin of all the faculties at work in cognition. Genealogy was thus central to Kant's analysis of the history of reason, but the organic affinity of reason would also be the key to understanding the text of the transcendental deduction itself.

When introducing the topic of affinity earlier in the transcendental deduction, Kant had opened with a series of questions: "As regards the empirical rule of association . . . upon what, I ask, does this rule, as a law of nature rest? How is this association possible? The ground of the possibility of the association of the manifold, so far as it lies in the object, is named the *affinity* of the manifold. I therefore ask, how are we going to make comprehensible to ourselves the thoroughgoing affinity of appearances, whereby they stand and *must* stand under unchanging laws?" (A113). The only way to make the affinity of appearances comprehensible, Kant responded, was by realizing that they had been grounded by a transcendental affinity within cognition itself. Only this kind of transcendental ground could guarantee, for Kant, that "all appearances stand in thoroughgoing connection according to necessary laws, and therefore in a transcendental affinity, of which the empirical is a mere consequence" (A114). Hume's mistake in appealing to the laws of association, as Kant later explained in the "Doctrine of Method," was that he had wrongly inferred "from the contingency of our determination *in accordance with the law*, the contingency of the *law* itself" and that he had thereby confounded "a principle of affinity, which has its seat in the understanding and affirms necessary connection, with a rule of association, which exists only in the imitative faculty of imagination, and which can exhibit only contingent, not objective connections" (A766/B794; cf. 4:259).[281] It was transcendental affinity, a principle "which has its seat in the understanding," that grounded the experience of nature's coherence; it was transcendental affinity that generated the empirical as its "mere consequence." As Kant also put it, the "objective ground of all association of appearances I entitle their *affinity*. And it is nowhere to be found save in the principle of the unity of apperception, in respect of all knowledge which is to belong to me" (A122).

While the skeptic might attempt to provide an empirical deduction of concepts taken from experience—a deduction appealing to the laws of association and grounded therefore upon the supposition of natural affinity—the most to be hoped for from such a deduction was the mere "semblance of conviction" regarding experience. This was an important mistake, as Kant saw it, since "the semblance of conviction which rests upon subjective causes of association and which is regarded as insight into a natural affinity, cannot balance the misgivings to which so hazardous a course must rightly give rise" (A783/B811). The experience of natural affinity was indeed grounded by acts of cognition but not in the contingent manner in which Hume had proposed it by way of the imagination and its laws of association. Natural affinity was produced by the organic unity of reason—that is, the genealogical affinity existing between reason and its diverse faculties engaged in the activities of cognition.

By appealing to "affinity" as a term meant to capture both the unity of cognition and the necessary unity and coherence of experience, Kant had chosen the perfect word for his purposes. For *affinity* perfectly captured the double meaning of not only a familial relationship existing between the faculties of cognition on the basis of their common descent from reason but also a structural resemblance, and thereby a necessary connection, between the "parts" of experience that had been constructed by the work of the faculties themselves. Transcendental affinity, in other words, lay at the heart of the deduction because it was an expression not only of organic unity but of Kant's transcendental theory of truth, a theory that had been predicated from the beginning on the fact that the means for making a logical connection between subjects and predicates were the same for connecting concepts and objects. Necessity could be guaranteed and the skeptic refuted, according to Kant, but it required showing that all of the tools relied upon by the empiricist—the abstraction of sensible concepts, the laws of association, and the experience of natural affinity—were in fact dependent upon the prior work of cognition. The organic unity of reason thus secured the possibility of cognition, just as transcendental affinity secured the possibility of coherent experience and thereby truth.

With transcendental and organic affinity finally in view, we can make sense of Kant's work throughout the *Critique of Pure Reason* to provide information regarding the "birthplace" of the various faculties responsible for cognition. Kant had introduced himself at the outset of the transcendental analytic, for example, as someone who had been engaged in "the hitherto rarely attempted *dissection of the faculty of understanding* itself," a dissection that had allowed him to investigate "the possibility of concepts *a priori* by looking for them in the understanding alone, as their birthplace." Kant's project would be an investigation into the origin of the categories, and thereby also a discovery of the *quid facti*, as the required first stage of the transcendental deduction. In pursuit of this, Kant explained that he had followed "the pure concepts to their first germs [*Keimen*] and dispositions [*Anlagen*] in the human understanding, in which they lie prepared, till at last, on the occasion of experience they are developed [*entwickelt*]" (A66/B91, cf. 4:274). What this investigation had revealed to him was that the table of judgments was the birthplace of the categories and that it was the judgments themselves that had served as the original "germs and predispositions" from out of which the categories had developed. Rehearsing conclusions that he had reached as early as 1773 regarding the identity of the logical forms of judgment and the intellectual concepts, Kant was ready to provide the "certificate of birth" that he sought (17:620). As he summarized it, "The same function which gives unity to the various representations *in a judgment* also gives unity to the mere synthesis of various representations *in an intuition*; and this unity, in its most general expression, we entitle the pure concept of the understanding" (A79/B104; cf. B159). And just so as not to leave the impression that the a priori origin of space and time as the forms of intuition might remain suspect until some kind of similar birth certificate be produced, Kant immediately singled out the case of space as presenting an obvious exception to the need for such an investigation into origin. "Geometry," Kant explained, "proceeds with security in knowledge that is completely *a priori*, and has no need to beseech philosophy for any certificate of the pure and legitimate descent of its fundamental concept of space" (A87/B120).

This exception granted to the forms of intuition would not be extended to the concepts of reason, and Kant in fact followed a strategy for tracing their lineage that was parallel to the one that he had employed with respect to the concepts of the understanding. Explaining that reason "itself gives birth to concepts," Kant suggested that by "following the analogy of concepts of understanding, we may expect that the logical concept will provide the key to the transcendental, and that the table of the functions of the former will at once give us the genealogical tree of the concepts of reason" (A299/B355–356). In the same way, therefore, that Kant had shown that the logical table of judgments gave rise to the concepts when the judgments were applied to sensible intuition, Kant would next argue that logical inferences could be discovered as the point of origin for the ideas of reason (A321/B378). In each of these cases, in the case of the categories of experience as much as that of the concept of reason, Kant appealed to logic because it could provide a "genealogical tree" with respect to the question of origin, and it was therefore critical for Kant to trace this ancestry given the centrality of transcendental affinity to his argument.

By locating the birthplace of both concepts and ideas in the table of judgments, however, Kant did not mean thereby to suggest that logic served as the ultimate ground of cognition. Kant had been clear since 1776 that reason had to serve as a source of unity that could independently ground the unity and coherence of experience provided by the rules for cognition (17:711). Even judgment, therefore, would have to be based upon reason as a prior ground of its unity. As Kant made the point, the transcendental unity of apperception "itself contains the ground of the unity of diverse concepts in judgment, and therefore of the possibility of the understanding, even as regards its logical employment" (B131). In fact, "The synthetic unity of apperception is therefore that highest point, to which we must ascribe all employment of the understanding, even the whole of logic, and conformably therewith, transcendental philosophy" (B134).

The unity of apperception or pure reason was thus the root of every branch of knowledge so far as cognition was concerned: it could be identified as the organic ground of space and time as the a priori forms of intuition (B161), and insofar as it grounded the logical table of judgments, it served as the ultimate basis of not only the concepts of the understanding but the ideas of reason as well. And all of this had to be shown in order to develop and support the argument for transcendental affinity standing at the center of the transcendental deduction. Only once intellectual concepts and the ideas of reason could be traced back to their birthplace in reason, only after reason could itself be identified as "self-born" and containing the "germ of its self-development," only then would knowledge be secured and the dogmatist and skeptic alike refuted. And this was all indeed a very long way for readers to travel when it came to discovering the underlying ground of Kant's argument in the transcendental deduction.[282]

When Kant took the opportunity to rewrite the transcendental deduction for the second edition of the *Critique of Pure Reason* in 1787, he was determined to respond to complaints regarding the obscurity of his argument. Thus although he would announce in his preface to the second edition that this version of the *Critique* offered relatively minor changes in exposition (Bxlii), in the case of the transcendental deduction Kant had in fact started over from scratch. Such rewriting allowed him to conclude the new piece with a taxonomical presentation of the systems of knowledge, a section mirroring, therefore, the

"History of Reason" discussion with which he had closed the book as a whole. Given the prominence of its location, however, what the added taxonomy marked was a renewed effort on Kant's part to identify with greater clarity the role played by the organic unity of reason for securing the proof that the deduction was supposed to supply.

Enlisting terms from his 1772 letter to Herz (10:131, A92/B124), Kant began his new conclusion—"Outcome of This Deduction of the Concepts of the Understanding"—with a reminder to his readers that there were "only two ways in which we can account for a *necessary* agreement of experience with the concepts of its objects: either experience makes these concepts possible or these concepts make experience possible" (B167). Taking up the two ways in turn, Kant explained that appealing to experience for one's concepts was like resorting to "a sort of *generatio aequivoca*," since necessity required the kind of transcendental grounds that could never emerge from something so passive as sense data alone. He continued: "There remains, therefore, only the second supposition—which is at the same time a system of the *epigenesis* of reason—namely, that the categories contain on the side of the understanding, the grounds of the possibility of all experience in general" (B167). This second supposition was of course Kant's position, but just as he had taken transcendental affinity to offer the best solution between Hume's assumption of a natural affinity and the innatists' belief in special affinity, Kant added innatism to his discussion, suggesting that it proposed a "middle course" between Hume and himself. An innatist such as Leibniz had rejected the empiricists' appeal to experience as much as Kant had when it came to necessary concepts, but instead of recognizing with Kant that the concepts were "*self-thought [selbstgedachte]* first principles *a priori* of our knowledge," Leibniz took them instead to be "subjective dispositions of thought, implanted in us from the first moments of our existence, and so ordered by our Creator that their employment is in complete harmony with the laws of nature in accordance with which experience proceeds—a kind of *preformation-system* of pure reason" (B167). The main problem with this approach, Kant argued, was not that it degraded the status of the concepts. The problem with innatism was that it degraded the quality of knowledge achieved via the categories altogether, since their divine origin eliminated the need for anything like a transcendental proof of their necessary connection to experience as the ground of its coherence. This, on Kant's view, made knowledge of experience wholly contingent, since the divine fiat by which its harmony had been originally granted could at any point be just as willfully revoked (B168; cf. 4:476). Only "the epigenesis of reason," as appealing neither to experience nor to God but only to itself, could finally serve as the true ground of experience. Only this view of reason would allow Kant, moreover, to describe reason as the author of the laws of nature with respect to what *was* the case and, as he would go on to show, as the author of the laws of freedom with respect to what *ought* to be so.[283]

When Kant sat down to write the appendix he had planned for the second edition of the *Critique*, he picked up directly from the point where he had left off in the architectonic—starting the piece, therefore, where the *Critique* had ended. Appealing to length considerations, however, Kant's publisher wisely rejected the proposed addition. The planned appendix thus appeared one year later instead as the *Critique of Practical Reason*. Kant began the newly detached second *Critique* just as he had when it was still an appendix, namely, with the demand that attention be paid to the unity of reason. Only by paying attention to this architectonic character of reason, Kant argued, would one be able to grasp the system of reason as a whole and see thereby the underlying unity of its parts. This included seeing more than the organic identity of theoretical and practical reason, however; it meant recognizing the methodological path that had had to be taken in order to follow the course of reason's development.[284] As Kant put it,

When it is a question of determining the origin, contents, and limits of a particular faculty of the human mind, the nature of human knowledge makes it impossible to do otherwise than begin with an exact and (as far as is allowed by the knowledge we have already gained) complete delineation of its parts. But still another thing must be attended to which is of a more philosophical and architectonic character. It is to grasp correctly the idea of the whole, and then to see all those parts in their reciprocal interrelations, in the light of their derivation from the concept of the whole, and as united in a pure rational faculty. (5:10)

Only once the whole system of reason had been grasped in this manner, according to Kant, could one move to the "second stage" of the investigation, that is, the "synoptic view, which is a synthetic return to that which was previously given only analytically" (5:10). Tracing a path set by reason itself, the *Critique of Pure Reason* had followed the course of reason's own developmental trajectory, offering up its *Bauplan* or "synoptic view" only once the parts of the system had been analytically presented and the whole could come into view.

CHAPTER SEVEN

Organic Logic: A Cautionary Tale

The *Critique of Pure Reason* was itself, therefore, a text whose own internal organization was patterned on reason's organic form. The linear progression of the book when read from the beginning mirrored a genetic approach to the gradual building up of experience from sensible intuition to intellectual concept to rational idea. This incremental course of development contained signposts throughout the text, however, announcements regarding the possibility of viewing the "bottom up" progress of Kant's discussion as something that could be viewed also as having been organized from the "top down." From the vantage point of the latter view, the pathway was reversed: it was the unity of reason that made the functioning of understanding possible, and it was the transcendental work of figurative synthesis that explained the work of intuition. Kant's passing indications of a potential shift in perspective came fully into view only at the end of the book, however, in his discussion of methodology. This account of the *Bauplan* of reason revealed the nonlinear structure of Kant's text, providing a map for reading the topography of the *Critique* as if it were an organic structure, one whose parts were not only interconnected such that a reader could view them equally from the "bottom up" or the "top down" but whose internal logic was reflexive, a teleological progression toward an end that had determined the fate of its path from the start.[285] To put it more simply, the course of the argument introduced in the "Doctrine of Method" was meant by Kant to force his readers to return to the beginning of the text, to see from the vantage point of an architect just how the structure would be progressively put together. The fate of reason and the intended course of its self-development would be mirrored in the manner by which Kant intended to lead his readers to an end that would return them to the beginning of his text with an idea of the whole in view. What this all meant for Kant was that he would have to initially present the telic course of reason's investigation as a linear account and to do so, moreover, in a manner that would make the realization of reason's prescribed end only possible by means of that linear advance. An acorn, to put this by way of analogy, would certainly grow up to be an oak under the right conditions. But its path from acorn to oak is not only linear, it is necessarily linear if it is ever to achieve the end toward which it was destined from the start. The structure of the *Critique of Pure Reason* was a working demonstration of Kant's commitment, in other words, to the idea that arguments too could be organically presented.

Recognizing this reflexive aspect to Kant's project allows one to see finally the means by which Kant sought to connect the two types of logic at work in his account. The transcendental logic of the understanding was discursive, it operated by means of the logic of conceptual determination, and it prescribed conditions that had been established on the backbone of the table of judgments. It was in this manner that the connection of concept and intuition could also be a connection of subject and predicate in the formation of a synthetic a priori judgment of experience. But insofar as transcendental logic ultimately required the organic unity of reason or transcendental affinity for its success, it was itself dependent upon an organic logic, one that was operating with entirely separate concerns. Organic logic was nondiscursive in its operations and nonlinear in its progression. This logic was modeled instead on organic cycles of generation and growth; it was dominated by genealogical concerns regarding lineage and affinity, and it made use, whenever it appeared, of the vocabularies of life: root, stem, branch, and birth. It was by means of this kind of logic that Kant had tried to make sense of the system of reason on the model of an organic whole, and it was only with this organic system in mind that he had been able to connect a critical theory of reason to the course of its natural history.[286]

What should not be forgotten amid conclusions like these, however, is the great difference between Kant's treatment of reason and the approach he took toward nature. Thus while Kant might have had the reflexive laws of an organic logic in mind when describing the work of reason, he did not believe that we could make anything like an identical claim regarding the laws by which an actual organic being might work. Similarly, although Kant thought it was reasonable to choose from organic models of generation when describing the epigenesis of reason, he would never have suggested that such a model was actually at work in the generation of natural organisms. Indeed, we might reflexively view a natural organism to be something that is both cause and effect of itself, but this attitude, according to Kant, could only arise on the basis of an analogy with our own reason, which truly was the cause and effect of itself. What one could say about reason, in other words, was entirely different from what one could say about the natural world.[287]

With these words of caution in mind, it is time to briefly look at what Kant had to say about organic life itself. In the *Critique of Pure Reason* Kant's discussion of organic life was oriented by taxonomical concerns, and he took his starting point from Plato. This was because Kant found Plato's theory of ideas useful for discerning not only taxonomical order but lines of natural affinity within the natural realm. Thus Kant explained, "It is not only where human reason exhibits genuine causality, and where ideas are operative causes (of actions and their objects), namely, in the moral sphere but also in regard to nature itself, that Plato rightly discerns clear proofs of an origin from ideas." As he developed the point, "A plant, an animal, the orderly arrangement of the cosmos—presumably therefore the entire natural world—clearly show that they are possible only according to ideas, and that though no single creature coincides with the idea of what is most perfect in its kind . . . these ideas are nonetheless completely determined in the Supreme Understanding, each as an individual and each as unchangeable, and are the original cause of things" (A317–18/B374). Kant's appeal to the "Supreme Understanding" in the creation of fixed species lines referred back to his search in 1763 for lines capable of active cycles of reproduction following an "initial divine organization." But his reference to Plato demonstrated the progression of Kant's thinking since then with respect to the special role of reason for introducing ideas regarding the unity of nature in its internal arrangement. Thus while he chided Leibniz, and Bonnet after him, for proposing a great chain of being within nature, Kant recognized that such a belief grew out of a "principle of affinity which rests on the interest of reason" and that served legitimate ends as a regulative principle for discovering lines of natural affinity within nature. "In this regulative capacity," Kant explained, reason "goes far beyond what experience or observation can verify; and though not itself determining anything, yet serves to mark out the path towards systematic unity" (A668/B696; cf. A660/B688). This caveat regarding reason as "not itself determining anything" positive with respect to its ideas of nature stood in stark contrast to practical reason's causality in producing ideas with actual efficacy in the moral sphere—this was the great "enigma" of reason, as Kant had put it in the second *Critique* (5:5). In the *Critique of Pure Reason*, both aspects of reason fell under the rubric of its *regulative* employment, but when Kant took up the task of thinking about nature and morality again in the *Critique of Judgment* he divided the labor, tasking "theoretical reflective judgment" with the problem of thinking nature's unity and putting "practical reflective judgment" in charge of moral teleology.[288]

The approach that "theoretical reflective judgment" would take with respect to its task was teleological. This was key to its ability to see "natural products" as more than simple machines, indeed as living beings, beings whose successful flourishing suggested they operated according to purposes set by the organisms themselves. Kant supplied a definition of this early in his discussion of "Teleological Judgment": "I would say, provisionally, that a thing exists as a natural purpose if it is *both cause and effect of itself* (although in two different senses). For this involves a causality which is such that we cannot connect it with the mere concept of nature without [also] regarding nature as acting from a purpose" (5:370–371). Putting aside an investigation into the two senses in which organisms could be viewed as cause and effect of themselves for the moment, Kant first outlined certain observable signs of purposiveness. Choosing a tree for his example, Kant explained that purposiveness was revealed in the tree's ability to preserve its species line through reproduction, in the tree's constant self-maintenance and growth, and in the interconnected functioning of all of its parts. As for the two senses in which the tree could be viewed as both cause and effect of itself, here Kant introduced two means for thinking about causality itself. The first viewed a linear progression of causes and effects in nature by means of a concept of causality that was provided by the understanding. The second approach looked instead for a teleological series, for a progression in which final causes served also as first causes; this was the approach taken by means of a concept of free causality or purposiveness that had been provided by reason. When viewed through reason's concept of an organic *nexus finalis*, the interconnected functioning of the parts of the tree, for example, demonstrated that it was "both an *organized* and *self-organizing* being" (5:374), or as Kant also put it, a being in which "*everything is a purpose and reciprocally also a means*" (5:376). By combining the two approaches, Kant was able to say that in the tree "the connection of *efficient causes* could at the same time be judged to be a *causation through final causes*" (5:373), a double view that allowed theoretical reflective judgment to finally regard the tree as both cause and effect of itself.

The main point to remember regarding natural organisms for Kant, however, was that the possibility of viewing them as "both organized and self-organizing beings" was itself based on an analogy made with respect to reason's own existence as both cause and effect of itself. Because an organized natural purpose was inconceivable by way of an analogy to a mechanical product, in other words, the analogy had to rely on reason and the kind of demonstration of free causality that it provided in the moral sphere (5:375, 5:396). This meant that ascribing purposiveness to an organism was something that was done in the service of reason's own investigations and that purposiveness was ultimately an idea generated by reason for the sake of itself.

As for attempts to explain organic generation, a case that seemed to demand the possibility of nature's vitality, Kant tirelessly cautioned against the temptation to trespass the limits of what could be positively known. In terms recalling his 1763 treatment of the topic in *The Only Possible Basis of a Proof for the Existence of God* (2:113–115), Kant favored a theory of generation that could balance the need for maintaining the species lines against the possibility of an organic generation of individuals within them. In the *Critique of Judgment*, however, Kant could identify the theory he favored by name: it was the position put forward by supporters of epigenesis or "generic preformation." Kant liked the theory in 1790 for much the same reasons he had liked its outlines in 1763: epigenesis reduced an appeal to supernatural agency to a bare minimum, since it relied on God for only the original construction of the forms that the species lines would take, and it balanced a mechanical account of nutrition and growth with a teleological explanation of the organism's purposive development. And Kant singled out Blumenbach's notion of a *Bildungstrieb* for praise, precisely because it seemed to offer empirical evidence of the theory of generic preformation itself (5:424).[289] Nonetheless, Kant's tone of caution regarding the life sciences was unchanged. However convincing our intuitions regarding nature's organic capacities might be, however promising the advances made by the life sciences might seem, the operating principles of the organism would simply never be revealed in an empirical investigation. When reason saw organic activity in nature, according to Kant, what it was really looking at was itself.[290]

Kant found epigenesis to be attractive for thinking about reason because it opened up possibilities for thinking about reason as an organic system, as something that was self-developing and operating according to an organic logic. Epigenesis thus served as a resource for a *metaphysical* portrait of reason, even as it was denied determinate efficacy in the *physical* world of organisms. This was not merely a metaphorical appeal, since Kant's use of the organic model had a deep methodological impact when it came to the critical system; indeed the system itself was conceived as a result of this model as an organic unity whose telic course of development could be described as a natural history of reason. Thus the interpretive question regarding Kant's account is not how this epigenetic conception of reason is supposed to connect to the careful edifice of determinate knowledge that is its product. The answer to that question is actually on view everywhere in the critical project once one starts looking for it. The question is whether Kant had to position reason in such a way that it would have to displace God—whose message conveyed across the history of religion required the grounding of rational theology, and whose generative work was mirrored by reason within the unity of its system—if there were to be any hope of reason ever satisfying itself.

EPILOGUE

A Daring Adventure of Reason

It is easy to miss the epigraph Kant chose for the second edition of the *Critique of Pure Reason*. A long quotation in Latin, the passage was taken from Francis Bacon's preface to the *Instauratio Magna*, or "Great Renewal," an unfinished collection of philosophical writings by Bacon including the *New Organon*. In quoting Bacon, Kant must have appreciated Bacon's hope that the *Instauratio* be seen as "the foundation of human utility and dignity" and even more Bacon's claim to have put forth in his work "nothing infinite, and nothing beyond what is mortal, for in truth it [the *Instauratio*] prescribes only the end of infinite errors" (Bii). Apart from such sentiments, however, it also made sense that Kant would choose an epigraph from Bacon for the first *Critique*, given the particular nature of Kant's project. Bacon's *New Organon* had dealt with inductive logic, a logic he intended to be a replacement for the deductive logic of Aristotle's *Organon*. Kant's *Critique of Pure Reason* was meant to provide a new logic as well, a transcendental logic capable of moving beyond the merely analytic conclusions of syllogistic reasoning and capable, thereby, of securing the claims reached by way of induction. For these reasons Bacon made for an obvious choice. But what exactly were the means by which Kant had finally secured Bacon's inductive practices? The answer to this question immediately revealed to Kant's readers that the choice of Bacon—father of the "New Science"—was in fact a case of subversive appropriation on Kant's part and that what it

EPILOGUE: A DARING ADVENTURE OF REASON

announced, more than anything else, was his specific intention with respect to a redefinition of empirical science altogether.

Consider the cover of the *New Organon*. It portrays a ship returning to the Mediterranean through the pillars of Hercules with the motto "Many will travel and knowledge will be increased" emblazoned below. There are sea monsters on either side of the ship, a second tribute to the mythical power of Hercules, slayer of monsters and protector of mankind. According to legend, the pillars astride the Straits of Gibraltar were remnants from the time Hercules had pulled apart the mountains in order to connect the two seas, the Atlantic and the Mediterranean. Renaissance tradition held that the pillars bore the warning *"nec plus ultra,"* nothing further beyond—though Plato had long before suggested that the lost city of Atlantis might indeed be located beyond them—so ships remained within the confines of the Mediterranean as a result. The *New Organon*'s image of a ship returning in full sail from the Atlantic, unharmed by the sea monsters surrounding it and passing like Hercules through the boundaries separating the known from the new, was perfect for capturing Bacon's goals for the New Science. For it demonstrated that knowledge and discovery were possible if only one was willing to leave familiar shores, a point to be visited by Bacon again in the *New Atlantis* only a few years later.

Compare now the language by which Kant took leave of the sections devoted to the achievements of transcendental logic as he turned toward the uncharted and dangerous waters of the "transcendental dialectic." Explaining that he had finished exploring the territory of the pure understanding—an exploration that had entailed a complete survey of its parts and a measurement of its extent—Kant summarized the results of his survey: "this domain is an island, enclosed by nature itself within unalterable limits. It is the land of truth—enchanting name!—surrounded by a wide and stormy ocean, the native home of illusion, where many a fog bank and many a swiftly melting iceberg give the deceptive appearance of farther shores, deluding the adventurous seafarer ever anew with empty hopes" (A235/B294).[291] In Kant's lexicon here the "adventurous seafarer" was reason. The practices of inductive science could be performed only in the land of truth and, indeed, as Kant saw it, Bacon's ships would require mooring altogether if there were to be any hope for the empirical sciences at all. This domain, this "land of truth," might be only an island, but according to Kant it was a land containing everything required for science to succeed against dogmatism and skepticism both. The physical sciences could rest secure in the knowledge that transcendental logic had undergirded the

FIGURE 1. A ship of discovery returning through the pillars of Hercules, frontispiece to Francis Bacon's *New Organon* included in his fragmentary *Instauratio Magna* (1620). The inscription reads: "Many will travel and knowledge will be increased" (Daniel 12:4).

world of experience, a landscape wherein cognition gave the law to nature (A159/B198). It was, contra Bacon, the attempt to leave the land of truth, to book passage with the "adventurous seafarer" in search of objects whose possession were forbidden to the human mind, that caused difficulties for the investigator of nature. It was no accident, therefore, that Kant had chosen to quote Persius in the opening pages of the *Critique*: "Dwell in your own house and you will know how simple your possessions are" (Axx).

These were, of course, all words that went unheeded by an "adventurous and self-reliant reason" (A850/B878), an adventurer who was naturally disposed, indeed fated (Avii) to move beyond the boundaries of experience when looking for something more than a mechanical explanation of nature's unity. From reason's vantage point, the observation of morphological similarities between organic forms suggested that lines of natural affinities connected the web of organic life. It was an irresistible conclusion and the basis for just the sort of investigation that reason was disposed to follow. Kant took the task of the transcendental dialectic, therefore, to be the exposure of such conclusions for what they were: ideas that had been projected by reason onto nature in the first place. Thus with respect to the question of natural affinity, Kant explained that while it was a principle of general logic that the "various species must be regarded merely as different determinations of a few genera, and these, in turn of still higher genera" (A651/B680), it was only after reason had projected this principle onto nature that its genealogical investigation at the empirical level had in fact begun. Reason's idea or "transcendental principle" regarding genera was not inherent to nature; its source was reason itself, and it rose out of the unity of reason in its search for a correlative unity in nature, in reason's search, in other words, for a mirroring of its *own* organic affinity such that "even amidst the utmost manifoldness" of nature it could "observe homogeneity in the gradual transition from one species to another, and thus recognize a relationship of the different branches, as all springing from the same stem" (A660/B688). It was Kant's task, in light of this, to repeatedly remind reason that it was the author of such principles and to protect it, as in this instance, from believing that it had discovered the basis of nature's unity in nature itself. This kind of belief was indeed the hallmark of what Kant took to be a "transcendental illusion" on reason's part, and it was the history of such illusions, so far as Kant saw it, that had made it impossible to establish metaphysics as a science up until the *Critique of Pure Reason* had appeared (A663/B691, A669/B697–A671/B699).

EPILOGUE

Kant returned once more to the topic of natural unity in the *Critique of Judgment* and his account showed that he had continued, in the intervening years, to keep abreast of the latest developments in natural history. In the *Critique of Judgment* Kant acknowledged the attractiveness of particular scientific hypotheses while insisting again that these hypotheses be recognized as arising out of reason for reason's own use in its investigation of nature. Kant could readily appreciate the attraction of comparative anatomy for natural historians, remarking that it was right to look with the "archaeologist" of nature for the means by which "a common archetype" could have connected the parts of nature through a principle of affinity. The idea of an archetype could be reasonably maintained insofar as "so many genera of animals share a certain common schema on which not only their bone structure but also the arrangement of their other parts seems to be based; the basic outline is admirably simple but yet was able to produce this great diversity of species, by shortening some parts and lengthening others, by the enfolding [*Einwickelung*] of some and the unfolding [*Auswickelung*] of others" (5:418). It was reasonable for naturalists to suspect that all species were "actually akin, produced by a common original mother," a position that Kant in fact considered more plausible than counterefforts to explain the emergence of species "by the mechanics of crude, unorganized matter." Indeed, by locating the origin of species in an organic "mother," naturalists were simply demonstrating reason's own need to suppose an organized basis for the production of organized beings. It was in this manner, as Kant put it, that the archaeologist of nature made

> mother earth (like a large animal) emerge from her state of chaos and make her lap promptly give birth to creatures of initially a less purposive form, with these then giving birth to others that became better adapted to their place of origin and their relations to one another, until in the end this womb itself rigidified, ossified, and confined itself to bearing definite species that would no longer degenerate [*ausarten*], so that the diversity remained as it had turned out when that fertile formative force [*fruchtbaren Bildungskraft*] ceased to operate. (5:419)

The naturalist proposing this scheme, according to Kant, was employing a teleological lens when viewing the history of the species lines, an idea of natural affinity that had been born out of a "daring adventure of reason." It was reason that had sought out the original point from which species had emerged and begun their gradual process of differentiation and it was reason that had demanded that the archaeologist "attribute to this universal mother an organization that purposively aimed at these creatures, since otherwise it is quite inconceivable how the purposive form is possible that we find in the products of the animal and plant kingdoms" (5:419). The search for final causes could not be eliminated from the study of nature, Kant argued, but it was a search that both began and ended with a principle rooted in reason alone.

Kant's stipulations regarding the heuristic as opposed to the constitutive use of principles supplied by reason would be unheeded by the majority of his philosophical and scientific contemporaries and successors. But the historical significance of reason's principle of natural affinity is particularly worth tracing, since it reveals the peculiar legacy of Kant's own turn to biological investigations when accounting for reason. What Kant had written regarding the natural history of reason—a system describing the means for approaching reason both as an individual and as a species with a history of transformations in its wake—was exactly the kind of account contemporary naturalists sought in their own investigations of organic life. And while the story of how Kant's synthesis of biological and epistemic considerations would be subsequently taken up and transformed by naturalists is too lengthy to be included at the end of a history of Kant's own development, one aspect of this legacy seems to require at least a brief remark.

One of the more striking aspects of Kant's appropriation by the life sciences was the way biological and epistemic considerations were fused in the use Goethe and Darwin would each make of Kant's transcendental principle of affinity. Goethe referred to Kant on numerous occasions throughout his various scientific studies and reflections. And he had been influenced by the *Critique of Judgment* in particular, even characterizing his own search for nature's archetypal forms as a case of having embarked on a deliberate "adventure of reason."[292] Darwin, by contrast, rarely mentioned Kant apart from passing references to his theory of race, but Kant's influence on Darwin by way of Goethe and the transcendental morphologists after him was still evident. For a principle of affinity worked within the systems of both of these naturalists not only as an epistemic principle for interpreting nature but as indeed a biological principle for connecting nature itself.

For Goethe, it was the archetype and its metamorphosis that would be responsible for both explaining the parts of nature and physically uniting nature as a whole. In his account of the archetype, Goethe described a point of orientation by which the naturalist could see the universal made individual insofar as the idea of the whole was always contained in the part. The archetype referred to an idea that was meant to retrain the eye, to teach it to search for the identity of the ideal and the real on view everywhere within nature. But unlike Kant's account of the principle of affinity, Goethe's approach to the archetype was not simply a matter of reflective judgments regarding the unity of nature. The metamorphosis of the archetype described the biological means by which the *actual* affinity of nature had been generated.

It was a move that would be strikingly similar to the one Darwin would make in the *Origin of Species* some half century later. Darwin's idea regarding the "common descent" of all species from one origin—Kant's womb of nature, so to speak—opened up an epistemic framework for interpreting nature on all its levels. It taught naturalists the proper approach to comparative anatomy, for example, one revealing this demonstration of affinity to be the true "soul of natural history."[293] But the idea of common descent was meant to do more than operate as the key concept for Darwin's so-called hypothetico-deductive approach. Descent with modification, as an idea for interpreting the interconnection of nature, was also intended as a description of the organic means by which such interconnection had occurred. The idea of common descent thus functioned like Goethe's archetype, as an epistemic means for the investigation of natural affinity at the same time that it was the biologically real ground of that affinity.

What these naturalists had done, from Kant's perspective, was to take reason's transcendental principle of affinity and force it into double service as a constitutive principle as well. It was perhaps easier to remember with Goethe that the archetypal ideas served as the means for teaching the scientist to see the identity of the real and the ideal, an act Goethe termed "intuitive perception," but when reading Darwin one had to be continually reminded that the idea of natural affinity was something that had been added by the investigator as a means for connecting a merely accumulated set of empirical facts. As Darwin himself would repeatedly remind readers of the *Origin of Species*, the idea of common descent simply put the facts together in a more satisfying manner than did the reigning theory regarding God's special creation of the species lines.

The main point to keep in mind when considering these appropriations of Kant's transcendental principle of affinity, however, is not the technical issue of a constitutive versus heuristic employment of the ideas of reason. What should instead be kept in view are the noticeably different agendas of Kant and the actual practitioners of natural history. They might well have inherited Kant's model of biological and epistemic synthesis from his treatment of reason, but the internal direction of the synthesis was reversed: where the naturalists began with reason's interpretation of nature, Kant had in fact used nature as a model by which to interpret reason. In its most radical form, epigenesis offered a theory of generation that Kant found compelling as a model for interpreting reason, for approaching reason as an agent that was both cause and effect of itself. But it was precisely the radicality of this model that led investigators in Kant's day to ultimately decide that this form of epigenesis was untenable as an explanation of nature. The most plausible versions of biological epigenesis, on their view, required the introduction of a formative capacity, some explanation of the means by which form could be conveyed beyond the models of simple replication presented by either crystal formation or a "vegetative force." This was the impulse behind Maupertuis's addition of psychic characteristics to his organic molecules, it was the reason Buffon had appealed to internal molds, and it had caused Blumenbach to put a formative drive at the seat of all life.

Kant understood all of this perfectly well. But he also saw the opportunities presented for a metaphysical interpretation of reason according to the most radical, and supposedly least tenable, version of physical epigenesis. This version allowed Kant to think of reason as a creature of its own making, as something self-born yet containing germs and predispositions for the possibility of its completion within an organic system that had been generated by itself. Germs and predispositions were not physical things for Kant in this case, nor did they lie in reason in the manner of preformed ideas. They existed merely as potentialities, susceptibilities, a virtual set of possibilities that, given the right environment—a school such as the Philanthropinum perhaps—could be realized. But until that moment of realization, a moment in no way predictable within the life of the individual, the model of epigenesis allowed the openness of reason's possibilities to be maintained. Only epigenesis, for Kant, could ground the description of a human being as a being with "an aptitude for purposes generally, i.e., in a way that leaves that being free" (5:431). Kant's vision regarding moral teleology may have been well in line with traditional notions of the constraints required for the perfection of freedom, but Kant in fact conceived of such constraints as laws that had been issued from out of freedom itself. Only the most radical model of biological epigenesis could allow Kant to imagine the heart of the critical system as he did, to maintain the organic identity of freedom and law in reason, and to describe the subsequent emergence of lawful constraints—be they categories for understanding nature or postulates for determining freedom—as laws that had been generated by reason solely for the sake of its own needs and purposes.

In light of this, it is hard to imagine that Kant would not have appreciated the possibilities for thought opened up by today's discussions of "epigenetics" and "emergent properties" in the life sciences when describing the fluid processes of organic life. The least tenable model has at last become the most plausible one for imagining the irreducible quality of the organism, one demanding our amazement not because of the intricate operations of its parts but because it has forced us to acknowledge the primacy of the living organic context within which such parts can emerge to mechanically function at all. This was precisely the kind of organic model that Kant had in mind when trying to grasp reason, and it is what locates him as a genuine forerunner of the organicism of both his day and our own.

Notes

INTRODUCTION

1. Schiller's description of the ideal society, for example, was of a political organization "which is formed by itself and for itself . . . insofar as the parts have been severally attuned to the idea of the whole," a state where "every individual enjoyed an independent existence, but could, when need arose, grow into the whole organism," *On the Aesthetic Education of Man*, trans. Reginald Snell (Mineola, NY: Dover, 2004), 33, 40. Herder was equally indebted to organic models: "Thus, as natural history can only observe a plant completely if it knows how it goes from seed, bud, bloom to decay, so would Greek history be for us such a plant." "General Reflections on the History of Greece," in Herder, *On World History: An Anthology*, ed. Hans Alder, trans. Ernest Menze (Armonk, NY: M. E. Sharpe, 1997), 288. Schlegel was also indebted to such models when arguing, for example, that "just as the organic seed [*Keim*]—thanks to the constant development [*Evolution*] of the formative drive [*Bildungstrieb*]—completes its cycle, grows vigorously, blossoms copiously, matures quickly, and wilts suddenly: so it is with every type of poetry, every age, every school of poetry." *On the Study of Greek Poetry*, trans. Stuart Barnett (Albany: SUNY Press, 2001), 65. And finally, in a vein similar to Schiller's, there is of course Kant's own remark on the body politic as analogous to the natural purposiveness of the organism. For in what seems to have been an allusion to the American Revolution, Kant wrote that "in speaking of the complete transformation of a large people into a state, which took place recently, the word *organization* was frequently and very aptly applied to the establishment of legal authori-

ties, etc., and even to the entire body politic. For each member in such a whole should indeed be not merely a means but also a purpose; and while each member contributes to making the whole possible, the idea of that whole should in turn determine the member's position and function." *Critique of Judgment*, 5:375. All citations from Kant will henceforth be to *Kants gesammelte Schriften* (Berlin: Walter de Gruyter, 1902–), except references to the *Critique of Pure Reason*, which will follow standard citation practice in referring to the A edition of 1781 and the B edition of 1787 when providing academy-edition page numbers.

2. Thomas Ramsay in praise of the naturalist Thomas Pennant, "To the Lovers of Natural History," *Scots Magazine* 34 (1774): 174.

3. The terms "preexistence" and "preformation" are frequently used interchangeably by commentators to capture the difference between a description of embryological formation where the problem of form is "solved" and a description, as in the case of epigenesis, where it is not. Jacques Roger, and Peter Bowler after him, have argued for the need to clearly distinguish between these terms. "Preexistence," as Roger sees it, should strictly refer to those theories proposing that all individual embryos were made by God at the moment of creation, so that all embryos thereby "preexist" their moment of specific temporal development. Malebranche, the earliest proponent of this view, argued that all future generations of the human race existed as fully formed miniscule beings whose embryological development was nothing more—so far as form was concerned—than their enlargement. Because Malebranche believed that future generations were contained in the sexual reservoirs of current ones, his position is referred to as *emboîtement*, the "Russian doll" theory, "encasement theory," and even "individual preformation." Initially, these miniscule "homunculi" were thought to be contained in the female, a position called "ovism"; once Leeuwenhoek discovered what he called "spermatic animalcules" under the microscope in 1674, the testes were thought instead to be the storage site, a determination that was referred to as "spermism." As positions like Malebranche's began to suffer under the pressure of discoveries such as Trembley's polyp, preexistence theories were adjusted until they became by the mid-eighteenth century, with Bonnet, arguments for the preexistence of only preformed germs for the species lines. "Preformation," according to Roger's distinction, should be reserved for a position like Buffon's. In this case, the parts of the embryo are formed by the parents (who contain molds for the parts, and whose molds were originally made for the species by God), with embryological development thus akin to the assembly of preformed parts. This account was disdained by preexistence theorists as affording too much power to nature, for even granting nature the capacity for the assembly of premade parts—Buffon thought this capacity was due to a "penetrating force"—was suspicious. Buffon insisted that generation was a mechanical process and has been since identified as a "me-

chanical epigenesist." The position that would be cautiously endorsed by Kant, proposed the nonmechanical (i.e., organic) epigenesis of individuals according to an internalized plan for their species as a whole, a plan that was therefore only "generic" for the species line. On the argument for severing preexistence and preformation, see Jacques Roger, *The Life Sciences in Eighteenth-Century French Thought*, trans. Robert Ellrich (Stanford, CA: Stanford University Press, 1997), 259–260; and Peter J. Bowler, "Preexistence and Preformation in the Seventeenth Century: A Brief Analysis," *Journal of the History of Biology* 4 (1971): 221–244. Against this distinction, see J. S. Wilkie, "Preformation and Epigenesis: A New Historical Treatment," *History of Science* 6 (1967): 138–150.

4. Aristotle, *Generation of Animals*, trans. A. L. Peck (Cambridge, MA: Harvard University Press, 1963), 733b23–735a29. Anthony Preus explicitly identifies Aristotle's account with epigenesis when discussing these passages. *Science and Philosophy in Aristotle's Biological Works* (Hildesheim: G. Olms, 1975), 66–69, 285n6.

5. William Harvey, *Disputations Touching the Generation of Animals* (1651), trans. Gweneth Witteridge (Oxford: Blackwell Scientific, 1981). For Aristotle's influence on Harvey see James Lennox, "The Comparative Study of Animal Development: William Harvey's Aristotelianism," in *The Problem of Animal Generation in Early Modern Philosophy*, ed. Justin E. H. Smith (Cambridge: Cambridge University Press, 2006), 21–46.

6. Caspar Friedrich Wolff, *Theoria generationis* (1759) and *Theorie von der Generation in zwo Abhandlungen erklärt und beweisen* (1764), facsimile reprints (Hildesheim: G. Olms, 1966). See Wolff (1759), §§ 43–53, §168, §242 and (1764), 160. The most thorough discussion of Wolff remains Shirley Roe's *Matter, Life, and Generation: Eighteenth-Century Embryology and the Haller-Wolff Debate* (Cambridge: Cambridge University Press, 1981). Karen Detlefsen challenges some of Roe's conclusions in "Explanation and Demonstration in the Haller-Wolff Debate," in Smith, *Problem of Animal Generation in Early Modern Philosophy*, 235–261.

7. Albrecht von Haller, "Reflections on the Theory of Generation of Mr. Buffon," in *From Natural History to the History of Nature: Readings from Buffon and His Critics*, ed. and trans. Phillip Sloan and John Lyon (Notre Dame, IN: University of Notre Dame Press, 1981), 322.

8. *Universal Natural History and Theory of the Heavens* (1755), 1:215–368.

9. *The Only Possible Basis for a Demonstration of the Existence of God* (1763), 2:63–164.

10. As Kant put the point three years later, "I am convinced that *Stahl*, who is disposed to explain animal processes in organic terms, was frequently closer to the truth than *Hoffman* or *Boerhaave*, to name but a few. These latter, ignoring immaterial forces, adhere to mechanical causes." *Dreams of a Spirit-Seer Elucidated by Dreams of Metaphysics* (1766), 2:331. The spirit of Kant's compromise between preformed species lines and organically

generated individuals can easily be compared to Ernst Mayr's own summary of the way genes work. As Mayr described it in 1997, "The genotype is the preformed element. But by directing the epigenetic development of the seemingly formless mass of the egg, it also played the role of the *vis essentialis* of epigenesis. . . . [The concept of a genetic program] was thus, in a way, a synthesis of epigenesis and preformation. The process of development, the unfolding phenotype, is epigenetic. However, development is also preformationist because the zygote contains an inherited genetic program that largely determines the phenotype." *This Is Biology: The Science of the Living World* (Cambridge: Cambridge University Press, 1997), 157–158. Mayr's formulation is nicely critiqued by Jason Robert, but this is a critique that Kant's sense of the *metaphysical* epigenesis of reason ultimately avoids given its affinity with Robert's understanding of epigenesis as entailing emergent properties. See Robert, *Embryology, Epigenesis, and Evolution: Taking Development Seriously* (Cambridge: Cambridge University Press, 2004), 38–41.

11. *Critique of Judgment* (1790), 5:165–486.
12. Kant is explicit regarding the spontaneous generation or "self-birth" (*Selbstgebärung*) of reason (A765/B793; cf. 18:273–275), but Harvey's model must be inferred when considering the relationship between the various faculties and apperception or reason as their undifferentiated ground (e.g., A119, B150–154).
13. Darwin famously described his response to reading the economist Thomas Malthus's *Essay on the Principle of Population* (1798) when developing the theory of natural selection with the comment, "Here, then, I had at last got a theory by which to work." *The Autobiography of Charles Darwin, 1809–1882, with Original Omissions Restored*, ed. Nora Barlow (London: Collins, 1958), 120. For Kant's use of epigenesis in this specific sense, see also 17:554, 18:8, 18:12, 18:273–275, B167. Compared to many of the issues under discussion in Kant scholarship, there has not been a great deal of work on Kant's appeal to epigenesis in connection with his account of reason; indeed the number of commentators can be counted on two hands. The best short essays remain Günter Zöller's "Kant on the Generation of Metaphysical Knowledge," in *Kant: Analysen-Probleme-Kritik*, ed. H. Oberer and G. Seel (Wurzburg: Königshausen and Neumann, 1988): 71–90; and Claude Piché's "The Precritical Use of the Metaphor of Epigenesis," in *New Essays on the Precritical Kant*, ed. Tom Rockmore (New York: Humanity Books, 2001), 182–200. Similarly significant for their attention to the distinctive philosophical requirements of the transcendental account are Hans Ingensiep's "Die biologischen Analogien und die enkenntnistheoretischen Alternativen in Kants Kritik der reinen Vernunft B §27," *Kant-Studien* 85, no. 4 (1994): 381–393; and Brandon W. Shaw's "Function and Epigenesis in Kant's *Critique of Pure Reason*" (master's thesis, University of Georgia, 2003). The most thorough discussion is provided by Thomas Haffner in

"Die Epigenesisanalogie in Kants Kritik der reinen Vernunft" (Ph.D. diss., Universität des Saarlandes, 1997). An older essay concentrating mainly on an explanation of the biological vocabulary used by Kant in the B deduction is provided by J. Wubnig, "The Epigenesis of Pure Reason: A Note on the *Critique of Pure Reason*, B, sec. 27, 165–168," *Kant-Studien* 60, no. 2 (1969): 147–152. A. C. Genova discusses the epigenesis of reason in the B deduction but primarily through the lens of Kant's later remarks regarding the epigenesis of organisms in the *Critique of Judgment*. See "Kant's Epigenesis of Pure Reason," *Kant-Studien* 65, no. 3 (1974): 259–273. The assumption that Kant's attitude toward epigenesis in biological organisms is the key to interpreting his account of the epigenesis of reason, is made by the majority of commentators, including Phillip Sloan's influential essay, "Preforming the Categories: Eighteenth-Century Generation Theory and the Biological Roots of Kant's A Priori," *Journal of the History of Philosophy* 40 (2002): 229–253; and John Zammito's several discussions indebted to Sloan's interpretation on this point, including most notably "'This Inscrutable *Principle* of an Original *Organization*': Epigenesis and 'Looseness of Fit' in Kant's Philosophy of Science," *Studies in History and Philosophy of Science* 34 (2003): 73–109. Ingensiep's response to the Sloan-Zammito interpretation is worth noting: "Organism, Epigenesis, and Life in Kant's Thinking," *Annals of the History and Philosophy of Biology* 11 (2006): esp. 70–73. Marcel Quarfood reaches different conclusions than Sloan and Zammito regarding Kant's supposed attitude toward preformation, but he follows the approach starting with Kant's biological discussions when considering the epigenesis of reason. See his *Transcendental Idealism and the Organism: Essays on Kant* (Stockholm: Almqvist and Wiksell International, 2004). This is also the case in Helmut Müller-Sievers's discussion of Kant in *Self-Generation: Biology, Philosophy, and Literature around 1800* (Stanford, CA: Stanford University Press, 1997); and in François Duchesneau's "Épigenèse de la Raison pure et analogies biologiques," in *Kant Actuel: Homage à Pierre Laberge*, ed. F. Duchesneau, G. Lafrance, and C. Piché (Montreal: Bellarmine, 2000): 233–256.

The difficulty with interpretations of the epigenesis of reason that begin with, or are at least oriented by, Kant's comments on biological generation is twofold. First, inadequate attention is typically given to the difference in status between transcendental and natural considerations, and thus to the specter of subreption regarding the latter. Second, the epistemic context of Kant's metaphysical appeal to the epigenesis of reason is frequently overlooked; that is, Kant's attempt to ground necessity in the face of Hume's challenge, on the one hand, and to locate the origin of knowledge in neither Locke's empiricism nor the innatism of Leibniz and Crusius, on the other. Ultimately Kant was a *metaphysician* with respect to reason, and because of this he was able to think about reason as something self-born even though he would have vigorously rejected the suggestion that he was thereby *naturalizing* reason in a vein similar (as he would have seen it) to the theory proposed by J. N. Tetens.

14. 17:736; cf. 28:231, 235, and 277, though this use is distinct from Kant's later identification of the transcendental imagination as a more fundamental ground of connection (23:18–20).

15. Kant later describes with approval Blumenbach's notion of a "formative impulse" (*Bildungstrieb*) so far as it refers to the capacity matter has for organization in the case of an organism (5:424). The formative impulse thus mirrors, as Kant interpreted it, the predisposition or aptitude of the mind for form.

16. In Bonnet's words, "I understand in general by the word germ every preordination, every preformation of parts capable by itself of determining the existence of a plant or of an animal." *The Contemplation of Nature* (1764), quoted by Bentley Glass in "Heredity and Variation in the Eighteenth Century Concept of the Species," in *Forerunners of Darwin, 1745–1859*, ed. Bentley Glass, Owsei Temkin, and William L. Strauss (Baltimore: Johns Hopkins Press, 1968), 167. Elizabeth Gasking offers a careful overview of Bonnet's views in *Investigations into Generation, 1651–1828* (London: Hutchinson, 1967), 117–129. For a recent reappraisal of Leibniz's influence on Bonnet's mature views, see François Duchesneau, "Charles Bonnet's Neo-Leibnizian Theory of Organic Bodies," in Smith, *Problem of Animal Generation in Early Modern Philosophy*, 285–314. While Bonnet was rightly famous for this "solution" to the problem of animal regeneration for preexistence theories, in his first essay on race Kant was more directly engaged with the work of Maupertuis and Buffon, each of whom also appealed to germs and dispositions in their discussion of the issues.

17. This term is used in current accounts of embryological development so far as the "epigenetic" response to environmental conditions understands actual ontogenesis as something that cannot be reduced to the simple unfolding of a genetic program. Emergent properties in this instance are "inexplicable from lower (or higher) hierarchical levels; for instance, cells' collective behavior during morphogenesis cannot be explained (or predicted) by examining the behavior of individual cells prior to cell division, differentiation, or (in animals) condensation—let alone by examining DNA sequences. This is because the formation of cell condensations is contingent not on genetic directives but rather on the spatiotemporal state of the organism and its component parts at multiple levels" (Robert, *Embryology, Epigenesis, and Evolution*, 97). A broad and relatively nontechnical recent assessment of emergent properties at work in cellular functioning is offered by Steve Talbott, "Getting Over the Code Delusion," *New Atlantis* 28 (2010): 3–27. Emergent properties in these discussions should not be confused with descriptions, for example, of appeals to "emergent vital forces" in late eighteenth-century German biology. In that context vital forces were understood to emerge from chemical-physical forces acting on

inorganic matter, and the question of a subsequently *directed* formation at the hands of these vital forces was either left open—as in the case of Caspar Wolff's *vis essentialis*—or included, as in Blumenbach's notion of the *Bildungstrieb*. A careful reconstruction of Wolff's position on this point is in François Duchesneau, "'Essential Force' and 'Formative Force': Models for Epigenesis in the Eighteenth Century," in *Self-Organization and Emergence in Life Sciences*, ed. B. Feltz, M. Crommelinck, and P. Goujon (Dordrecht: Springer, 2006), 171–186, esp. 173–175. Timothy Lenoir nicely describes Blumenbach's position, according to which "the *Bildungstrieb* was not a blind mechanical force of expansion which produced structure by being opposed in some way; it was not a chemical force of 'fermentation,' nor was it a soul superimposed on matter. Rather the *Bildungstrieb* was conceived as a teleological agent which had its antecedents ultimately in the inorganic realm but which was an emergent vital force." "Kant, Blumenbach, and Vital Materialism in German Biology," *Isis* 71 (1980): 83.

18. Leibniz describes this kind of special affinity in the *New Essays on Human Understanding* (1705), ed. and trans. Peter Remnant and Jonathan Bennett (Cambridge: Cambridge University Press, 1996), 80. Hume, on Kant's reading, "confounds a principle of affinity, which has its seat in the understanding and affirms necessary connection, with a rule of association, which exists only in the imitative faculty of imagination, and which can exhibit only contingent, not objective, connections" (A766/B794, cf. 4:259). See Hume, *A Treatise of Human Nature* (1739–1740), ed. L. A. Selby-Bigge, rev. ed., ed. P. H. Nidditch (Oxford: Oxford University Press, 1978), 504n71.

19. The "system of evolution" refers in this case to preexistence theory, specifically the encasement or *"emboîtement"* model of generation. For more discussion of this see n. 3.

20. An archaeology of nature attempting to link species, while not inconsistent as a judgment of reason, has no empirical evidence and therefore amounts to what Kant describes as a "daring adventure of reason" (5:419n1). Judgment regarding the unity of humanity as a result of their monogenesis has at least the evidence of interfertility in its support.

21. Johann Wolfgang von Goethe, "Fortunate Encounter" (1794), in *Scientific Studies*, trans. and ed. Douglas Miller, vol. 12 of *Goethe's Collected Works* (New York: Suhrkamp, 1983), 21. Cf. "What I had undertaken to do was nothing less than to present to the physical eye, step by step, a detailed, graphic, orderly version of what I had previously presented to the inner eye conceptually and in words alone, and to demonstrate to the exterior senses that the seed of this concept might easily and happily develop into a botanical tree of knowledge whose branches might shade the entire world." "Later Studies and Collections" (1817), in *Goethe's Botanical Writings*, trans. and ed. Bertha Mueller (Woodbridge, CT: Ox Bow, 1952), 97. Robert J. Richards highlights the shift from a heuristic to a constitutive use of Kant's

approach to nature in "Kant and Blumenbach on the *Bildungstrieb*: A Historical Misunderstanding," *Studies in the History and Philosophy of Biological and Biomedical Science* 31 (2000): 11–32. See also Richards's *The Romantic Conception of Life: Science and Philosophy in the Age of Goethe* (Chicago: University of Chicago Press, 2002), chap. 5., 216–237.

22. "This resemblance is often expressed by the term 'unity of type': or by saying that the several parts and organs in the different species of the class are homologous. The whole subject is included under the general name of Morphology. This is the most interesting department of natural history, and may be said to be its very soul." Darwin, *The Origin of Species by Means of Natural Selection; or, The Preservation of Favoured Races in the Struggle for Life* (1859) (London: Penguin Books, 1968), 415.

CHAPTER ONE

23. See, for example, "De Anima," 415b, 9–30, in *The Complete Works of Aristotle*, ed. Jonathan Barnes, vol. 1 (Princeton, NJ: Princeton University Press, 1984). For some discussion of the role played by metaphysics for Aristotle's theory of sexual reproduction, see J. M. Cooper, "Metaphysics in Aristotle's Embryology," *Proceedings of the Cambridge Philological Society* 214 (1988): 14–41; A. Code, "Soul as Efficient Cause in Aristotle's Embryology," *Philosophical Topics* 15 (1986): 51–60; and D. Henry, "Understanding Aristotle's Reproductive Hylomorphism," *Apeiron: A Journal of Ancient Philosophy and Science* 39 (2006): 269–300. While the situation is more complicated when explaining the spontaneous generation of lower animals, the metaphysical models are still presupposed, and in fact synonymy is preserved. A helpful discussion of this is in D. Henry, "Themistius and Spontaneous Generation in Aristotle's Metaphysics," *Oxford Studies in Ancient Philosophy* 24 (2003): 183–208.

24. In a typical formulation Jean Calvin declares that "concerning inanimate objects, we ought to hold that, although each one has by nature been endowed with its own property, yet it does not exercise its own power except insofar as it is directed by God's ever-present hand. These are, thus, nothing but instruments to which God continually imparts as much effectiveness as he wills, and according to his own purpose bends and turns them to either one action or another." *Institutes of the Christian Religion*, ed. J. McNeil, 2 vols. (Philadelphia: Westminster, 1960), bk. 1, chap. 16, sect. 2. A well-researched discussion of the impact of Reformation theology on seventeenth-century mechanical philosophy is Gary B. Deason's "Reformation Theology and the Mechanistic Conception of Nature," in *God and Nature: Historical Essays on the Encounter between Christianity and Science*, ed. David C. Lindberg and Ronald L. Numbers (Berkeley: University of California Press, 1986): 167–191. A clear account of Boyle's work to make sense of matter in motion within the constraints set by reformers

is in Peter Anstey's *The Philosophy of Robert Boyle* (New York: Routledge, 2000), esp. 164ff.
25. The classic example of this is Borelli's *De motu animalium* (1680–1681), but Descartes's *Treatise on Man* serves just as well. A survey of contributors to the rise in mechanist anatomy is in R. S. Westfall's "Biology and the Mechanical Philosophy," in *The Construction of Modern Science: Mechanisms and Mechanics* (Cambridge: Cambridge University Press, 1977): 82–104.
26. This formative work on the part of the soul is distinct from the role played by matter with respect to individuation. On the difference see G. E. R. Lloyd, "Aristotle's Principle of Individuation," *Mind* 79 (1970): 510–529.
27. "Parts of Animals," 643b27f., in *The Complete Works of Aristotle*, ed. Barnes. See also G. E. R. Lloyd's "The Development of Aristotle's Theory of the Classification of Animals," *Phronesis* 6 (1961): 59–81. Aristotle's caution was overlooked in the face of the overwhelming practical needs facing taxonomists in the sixteenth century. The most important figure in this history was Andreas Cesalpino. Cesalpino was determined to develop botany as a proper science, but to do so he had to retrieve it from the province of medical gardeners and their chaotic classification schemes within the many *materia medica* being produced at the time. In contrast to these sorts of practical aims regarding the development of medicinal recipes, Cesalpino's interests were primarily theoretical, and he saw the development of a universal classification system to be the necessary basis for any true botanical science. Taking his lead from Aristotle, Cesalpino argued that reproduction was the essential function of a plant and that a natural system of division could therefore be established according to the parts of fructification as the most essential features of a plant. As he put it, "From the means of producing fruits many genera of plants can be distinguished. Indeed, in no other structures has nature formed such a multiplicity and distinction of organs as are seen in the fruits. . . . Therefore we shall try to investigate the genera of plants by means of the unique fructifying characters which have been provided us by the Grace of God, both in the trees and shrubs, and in other plants." *De plantis libri XVI* (Florence, 1583), bk. I, 28. Phillip R. Sloan emphasizes Cesalpino's incorporation of Aristotle in "John Locke, John Ray, and the Problem of the Natural System," *Journal of the History of Biology* 5 (1972): 1–53, esp. 9–13. For further discussion of Cesalpino see Julius von Sachs's discussion in his *History of Botany (1530–1860)*, trans. Henry Garnsey (Oxford: Clarendon, 1906), 37–66; and A. G. Morton's *History of Botanical Science: An Account of the Development of Botany from Ancient Times to the Present Day* (London: Academic, 1981), 128–148.
28. "The Origin of Forms and Qualities According to the Corpuscular Philosophy," in *Selected Philosophical Papers of Robert Boyle*, ed. M. A. Stewart (Indianapolis: Hackett, 1991), 49ff. For a lengthier discussion see Boyle's 1675 essay, "Of the Imperfection of the Chemists' Doctrine of Qualities," ibid., 120–137.

NOTES TO CHAPTER 1

29. On this point Boyle's target was the Aristotelians' reliance on a substantial form to provide unity to matter and, in particular, Daniel Sennert's hybrid of corpuscular-Aristotelianism. See William Newman, *Atoms and Alchemy: Chymistry and the Experimental Origins of the Scientific Revolution* (Chicago: University of Chicago Press, 2006). Boyle's language of matter's convention is meant, therefore, to replace the metaphysical concepts of both substance and form, arguing, moreover, that discussions of generation, corruption, and alteration can be adequately redescribed in terms of matter's convention, dissolution, and transposition resulting from local motion. See Boyle's "The Origin of Forms and Qualities According to the Corpuscular Philosophy," in *Selected Philosophical Papers*, , 44. The convention or "stamp" of corpuscles can thus explain the relatively stable properties demonstrated by metals, for example, without compromising the basic ontology regarding matter's essential plasticity: "For such a convention of accidents is sufficient to perform the offices that are necessarily required in what men call a *form* since it makes the body such as it is, making it appertain to this or that determinate species of bodies, and discriminating it from all other species of bodies whatsoever." Ibid., 40. On this see also Dennis Des Chene, "From Natural Philosophy to Natural Science," in *The Cambridge Companion to Early Modern Philosophy*, ed. Donald Rutherford (Cambridge: Cambridge University Press, 2006): 67–94, 79.

30. Thus despite the fact that Boyle was the first person to develop chemical identification tests, for example, the ontological theory guiding Boyle's investigations meant that he was unable to discern their true significance for the development of a system of classification. See Richard S. Westfall's discussion in *Construction of Modern Science*, 79: "Again [Boyle's] mechanical philosophy appears to have operated to thwart the most promising aspect of his chemistry." It should be noted that recent work on Boyle's chemistry has suggested, against a long-standing tradition in line with Westfall's reading, that interpretation of Boyle's corpuscular ontology cannot simply understand it according to its mechanical principles but should in fact include the integration of *semina rerum*—particles endowed with different degrees of formative force and therefore not substantially identical—into corpuscular philosophy by the mid-1750s. See especially Antonio Clericuzio's discussion of Boyle in his *Elements, Principles, and Corpuscles: A Study of Atomism and Chemistry in the Seventeenth Century* (Dordrecht: Kluwer, 2000). A reconsideration of "inert matter" can also be found in Simon Schaffer, "Godly Men and the Mechanical Philosophers: Souls and Spirits in Restoration Natural Philosophy," *Science in Context* 1 (1987): 55–85; and John Henry, "Occult Qualities and the Experimental Philosophy: Active Principles in Pre-Newtonian Matter Theory," *History of Science* 24 (1986): 335–381.

31. "The Origin of Forms and Qualities According to the Corpuscular Philosophy," in *Selected Philosophical Papers*, 70.

32. "Considerations and Experiments, Touching the Origin of Qualities and Forms. The Historical Part," in *The Works of Robert Boyle*, ed. Michael Hunter and Edward B. Davis, 14 vols. (London: Pickering and Chatto, 1999), 5:383–384.
33. Boyle's recourse to a physical yet "plastick" principle when explaining generation demonstrates the genuine difficulties faced by midcentury theorists in accounting for biological processes. As Peter Anstey describes Boyle's position, "Study of Boyle's theory of seminal principles reveals a Boyle who is in tension, not a Boyle who abandons the corpuscular hypothesis when intruding on the biological domain and not a Boyle who is unaware of the need to reach beyond the sparse ontology of mechanical affections of matter. Boyle was unable to resolve this dilemma in his natural philosophy and as interpreters we should not do it for him." "Boyle on Seminal Principles," *Studies in the History and Philosophy of Biological and Biomedical Sciences* 33 (2002): 628.
34. "Considerations and Experiments," 384. Boyle's description of formative power in terms of a motion for fitting together particles is perhaps not so far from Descartes's discussion of bodily processes in his *Treatise on Man*; generation, for Descartes, is due to motion yielded by the heat of fermentation (like "yeast"), and this fermented mixing of the seminal fluids from the two sexes moves the individual particles into the form required to become parts of the body. *Description of the Human Body*, AT 253. Further discussion is in Vincent Aucante, "Descartes's Experimental Method and the Generation of Animals," in *The Problem of Animal Generation in Early Modern Philosophy*, ed. Justin H. Smith (Cambridge: Cambridge University Press, 2006): 65–79. The critical role played by motion and heat for Descartes and Boyle reveals the seventeenth century's pervasive indebtedness to Aristotelian models. See the helpful discussion of Aristotle's theories and influence in Remke Kruk, "A Frothy Bubble: Spontaneous Generation in the Medieval Islamic Tradition," *Journal of Semitic Studies* 35 (1990): 265–282, esp. n. 1.
35. See J. W. Gough, "John Locke's Herbarium," *Bodleian Library Record* 7 (1962–1967): 42–46; Peter Anstey and Stephen Harris, "Locke and Botany," *Studies in the History and Philosophy of Biological and Biomedical Science* 37 (2006): 151–171; G. G. Meynell, "A Database for John Locke's Medical Notebooks," *Medical History* 42 (1997): 473–486; J. R. Milton, "Locke, Medicine, and the Mechanical Philosophy," *British Journal for the History of Philosophy* 9 (2001): 221–243; and Guy Meynell, "Locke as a Pupil of Peter Stahl," *Locke Studies* 1 (2001): 221–227.
36. Locke's "Morbus" entry is reproduced in Jonathan Walmsley's "Morbus—Locke's Early Essay on Disease," *Early Science and Medicine* 5 (2000): 391–393; all citations are from p. 392, English modernized. Walmsley argues for the influence had by Van Helmont's philosophy on Locke's position here in contradistinction to Boyle's. See also J. R. Milton's discussion of Locke

and Van Helmont in this context, "Locke, Medicine, and the Mechanical Philosophy," *British Journal for the History of Philosophy* 9 (2001): 221-243. Walmsley's view is contested by Peter Anstey and subsequently rebutted by Walmsley. See Peter Anstey's "Robert Boyle and Locke's 'Morbus' Entry: A Reply to J. C. Walmsley," *Early Science and Medicine* 7 (2002): 358-377 and Jonathan Walmsley's "Morbus, Locke and Boyle: A Response to Peter Anstey," *Early Science and Medicine* 7 (2002): 378-397.

37. Locke makes use of neither "metamorphosis" nor "epigenesis" to describe the chicken's embryonic change from liquid to hard parts. Although Locke had carefully worked through Harvey's *De generatione*, taking care to note both Harvey's distinction between "Metamorphosis" and "Epigenesis" and his discussion of the efficient cause of generation (see Locke's medical school notebook entries from 1659-1660, Bodleian Library MS Locke fol. 14, p. 1; fol. 20, pp. 1-2, 4-5), it is not clear that Locke has this model in mind or even, pace Walmsley, is instead contrasting the Helmontian conception of a guiding "archeus" to Boyle's conception of motion guiding fitted particles of matter into the chick. J. R. Milton takes "Morbus" to represent an early eclecticism on Locke's part. See "Locke, Medicine, and the Mechanical Philosophy," 239.

38. In Milton, "Morbus—Locke's Early Essay on Disease," 392, English modernized.

39. Thomas Sydenham, preface to *Observationes Medicae*, trans. in G. G. Meynell, "Locke and the Preface to Sydenham's *Observationes Medicae*," *Medical History* 50 (2006): 106.

40. In "The Origin of Forms and Qualities According to the Corpuscular Philosophy" Boyle also appeals to mistletoe, in this case as an example against the supposed existence of a vegetative soul guiding the plant. *Selected Philosophical Papers*, 66.

41. Authorship of Sydenham's preface has been attributed to Sydenham, to Locke, and to both together. See Guy Meynell's "Locke and the Preface to Sydenham's *Observationes Medicae*," *Medical History* (2006): 93-110; and Milton, "Locke, Medicine, and the Mechanical Philosophy," esp. 229n42.

42. One complaint in the preface already speaks to the problem of classification so far as the *materia medica*—compendiums of medicinal recipes—are said to lack utility because of the inconsistent theories of symptoms and disease guiding their organization. See Meynell, "Locke and the Preface to Sydenham's *Observationes Medicae*," 103.

43. Locke to Dr. Thomas Molyneux, January 20, 1692, in *Dr. Thomas Sydenham (1624-1689)* by Kenneth Dewhurst (Berkeley: University of California Press, 1966), 179-180.

44. *Elements of Natural Philosophy*, in *The Works of John Locke*, 12th ed., (London: C. and J. Rivington, 1824), 3:319.

45. A large piece of Locke's second reply to Stillingfleet takes up the latter's use of Leeuwenhoek's discovery when discussing resurrection. Locke argues

that while seeds are responsible for both the production of individuals and the continuation of species, there can be no sense to the suggestion that the preformed individual in embryo is materially identical to the adult. "Locke's Reply to the Bishop of Worcester's Answer to his Second Letter," in *Works of John Locke*, 4:319). Locke's response flows directly from his discussion of identity added to the second edition of the *Essay*: it is the "organization of life in several successively fleeting particles of matter united to it" that makes for continued identity; to suppose that it were matter alone would make it hard "to make an *embryo*, one of years, mad, and sober, the same man". Locke, *An Essay Concerning Human Understanding*, ed. Niddich (Oxford: Clarendon Press, 1975), 2.27.3, 2.27.6. Although it has been conventional in Locke studies to cite the *Essay*'s book, chapter, and section divisions as, for example, *E* II. xxvi. 2, I will be following more recent trends in simply using Arabic numerals separated by periods to indicate the divisions.

46. Bodleian Library MS Locke fol. 2, 357–358.
47. J. W. Gough details the contents of Locke's *Herbarium* in "John Locke's *Herbarium*," 42–46. For an extensive discussion of the circumstances surrounding Locke's collection practices and his creation of the *Herbarium*, see Anstey and Harris, "Locke and Botany."
48. For example, "I think, there is scarce any one will allow this upright figure, so well known, to be the essential difference of the *Species Man*; and yet how far Men determine of the sorts of Animals, rather by their Shape, than Descent, is very visible; since it has been more than once debated, whether several human *Foetus* should be preserved, or received to Baptism, or no, only because of the difference of their outward Configuration, from the ordinary Make of Children, without knowing whether they were not as capable of Reason, as Infants cast in another Mould" (3.6.26).
49. See Locke's entry on "Species" from 1677: "in vegetables we find that several sorts come from the seeds of one and the same individual as much different species as those that are allowed to be so by philosophers." Locke's journal entry from November 19, 1677, in *An Early Draft of Locke's Essay, Together with Excerpts from His Journals*, ed. by R. I. Aaron and J. Gibb (Oxford: Clarendon Press, 1936), 99.
50. It must be said that Locke followed this description of the life of plants with an account favorably comparing the workings of animal parts to the functioning of a clock. One can only speculate as to the grounds for his greater openness to living aspects of botanical processes, but his own experience with these had at least prepared him to be ready for surprises when it came to vegetable life.
51. While imperceptible corpuscles are assumed to bear a causal relationship to the observed properties of any given thing, there is no sense in which these can be known, for not only are they insensible but our constitution prevents us from experiencing unsorted aggregates. We appreciate roses,

for example, for their fragrance and color and handle them gingerly both for fear of thorns and in deference to the delicacy of their petals; the experience of a rose, however, is in no way akin to that of a body conceived as a collection of corpuscles because the latter describes an experience one could never actually have. Locke does not seem to have perceived any tension between his appeal to an unknowable real essence and his account of the cognitive means by which experience was in fact constructed. But the relationship between these two views explains, I think, the appearance of apparent contradictions between claims, for example, that nature has a real constitution, and that the very notion of an internal constitution is incoherent insofar as it requires criteria for determining it.

52. Despite his talk of abstraction and archetypes, Locke was not an idealist, and his materialist commitments were typically on view in such discussions. That aside, this kind of tension between one's experience of real individuals and the simultaneous acknowledgment of the artificial nature of species categories, continues to plague discussions in natural history to this day.
53. Locke to William Molyneux, January 20, 1693, in *The Correspondence of John Locke*, ed. E. S. De Beer (Oxford: Clarendon, 1979) 4:626.
54. Cf. "Whereby it is plain, that Men follow not exactly the Patterns set them by Nature, when they make their general *Ideas* of substances; since there is no Body [such as referred to by "Metal"] to be found, which has barely Malleableness and Fusibility in it, without other qualities as inseparable as those. But Men, in making their general *Ideas*, seeking more the convenience of Language and quick dispatch, by short and comprehensive signs, than the true and precise Nature of Things, as they exist, have, in the framing their abstract *Ideas*, chiefly pursued that end, which was, to be furnished with store of general, and variously comprehensive Names (3.6.31). Locke's nominalism is at its most pronounced with respect to non-living substances so far as these seem to succumb to the demands of corpuscular ontology. See especially Locke's journal entry on "Species," September 19, 1676, in *An Early Draft of Locke's Essay*, 83.
55. Lisa Downing makes the point as well, arguing, for example, that an "unsorted particular" could not count as a real essence for Locke, "since no distinction between essential and accidental properties is possible without reference to a kind." "Locke's Ontology," in *The Cambridge Companion to Locke's Essay Concerning Human Understanding* ed. Lex Newman (Cambridge: Cambridge University Press, 2007). Anstey and Harris, by contrast, take Locke's active involvement in botanical matters to raise important questions against the presumption that Locke was a species nominalist: "Locke's botanical activities link him closely with essentialist classificatory projects, whilst his interpreters, using the *Essay* as their entry point into Locke's views on species, seem uniformly to have taken him to be, if not a species nominalist, then at least highly skeptical of the essentialist

program in biological classification in general." "Locke and Botany," 167. While they take the extended discussions of species in book 3 of the *Essay* to represent rather a "moderate conventionalism" (168), for the reasons argued above I agree with the stronger reading of Locke's nominalism, and I take it to be motivated in part as a result of precisely those empirical investigations Anstey and Harris see in their favor.

56. *Some Thoughts Concerning Education*, ed. R. H. Quick (Cambridge: Cambridge University Press, 1902), §192, 168. Reading this passage, G. A. J. Rogers concludes that "Locke's ontology, then, allowed room for spirits, and therefore appears to allow for the possibility of the spirits of the natural magicians," and he suggests, therefore, that "Locke's rejection of the possibility of knowledge of the essences of substances—material or spiritual—did not commit him either to the rejection of an ontology which could include active spirits, or, on the other side, to one that excluded the possible truth of Epicurean Atomism." *Locke's Enlightenment: Aspects of the Origin, Nature and Impact of his Philosophy* (Hildesheim: G. Olms Verlag, 1998), 185–186.

57. A very clear account of Leeuwenhoek's discovery and subsequent difficulties in explaining the precise manner by which the spermatic "animalcules" became active only at the age of sexual maturity and the subsequent physical process by which they accrued matter is detailed in Edward G. Ruestow, "Images and Ideas: Leeuwenhoek's Perception of the Spermatozoa," *Journal of the History of Biology* 16 (1983): 185–224.

58. While Leibniz's system would eventually furnish some theoretical foundations for discoveries made by the microscopists, as Shirley Roe rightly observes, "The fact that [Swammerdam] and Malpighi's observations were immediately taken up by those making preformationist claims indicates that the concept of preexistence was not one that grew out of observational evidence alone. It is clear, as several scholars have pointed out, that preformation through preexistence was a theory that responded more to philosophical than to observational needs." In Shirley Roe, *Matter, Life, and Generation* (Cambridge: Cambridge University Press, 1981), 7. Roe offers a brief overview of developments in generation theory from Harvey's epigenetic account to the various camps—ovism, animalculism, and Perrault's germ theory—within preexistence theory, 1–20. Far more detailed discussions can be found in the opening chapters of Elizabeth Gasking's *Investigations into Generation, 1651–1828* (London: Hutchinson and Co., 1967), and in Jacques Roger's *The Life Sciences in Eighteenth-Century French Thought*, translated by Robert Ellrich (Stanford: Stanford University Press, 1997), especially 205–369.

59. As Leibniz puts it in the *Principles of Nature and Grace based on Reason* (1714), "Each monad, together with a particular body, makes up a living substance. Thus, there is not only life everywhere, joined to limbs or organs, but there are also infinite degrees of life in the monads, some domi-

nating more or less over others." In *Philosophical Essays*, ed. and trans. R. Ariew and D. Garber (Indianapolis: Hackett, 1989), 208. The best treatment of Leibniz in this context is Justin E. H. Smith's *Divine Machines: Leibniz and the Sciences of Life* (Princeton, NJ: Princeton University Press, 2011).
60. Ibid., 222.
61. *A New System of the Nature and Communication of Substances, and of the Union of the Soul and Body* (1695), in *Philosophical Essays*, 139. "Only metaphysical points or points of substance (constituted by forms or souls) are exact and real," Leibniz continued, "and without them there would be nothing real, since without true unities there would be no multitude" or individuation at all. Ibid., 142. See also *Discourse on Metaphysics* (1686), ibid., 44.
62. Leibniz went so far in his support for the microscopists as to include a call for their support in his essay on justice. See his *Reflections on the Common Concept of Justice* (1702), in *Philosophical Papers and Letters*, ed. and trans. L. Loemker (Chicago: University of Chicago Press, 1956), 2:532.
63. Leibniz to Arnauld, October 9, 1687, ibid., 1:532.
64. Leibniz to Louis Bourguet, August 5, 1715, ibid., 2:1079.
65. Leibniz to Arnauld, October 9, 1687, ibid., 1:531. In 1687 part of Leibniz's concern, as the letter continues, is whether the individual monads make up the "organic machine." Smith carefully distinguishes the strands of Leibniz's approach to living beings in terms of their physical, organic, and metaphysical aspects in *Divine Machines*, 97–123.
66. In the *New System* Leibniz summarizes his argument for this. See *Philosophical Essays*, 140–141.
67. Leibniz considered the organized soul-body complex of an animal to operate teleologically at the metaphysical level and mechanically at the physical level (though this did not mean that he took derivative mechanical forces to be less purposive). His descriptions of the organic machine thus often presented a blend of mechanical, natural, and metaphysical elements, as in the following response to Locke's description of the understanding as akin to an empty closet: "To increase the resemblance we should have to postulate that there is a screen in this dark room to receive the species [sense ideas], and that it is not uniform but is diversified by folds representing items of innate knowledge; and, what is more, that this screen or membrane, being under tension, has a kind of elasticity or active force, and indeed that it acts (or reacts) in ways which are adapted both to past folds and to new ones coming from impressions of the species. This action would consist in certain vibrations or oscillations, like those we see when a cord under tension is plucked and gives off something of a musical sound." *New Essays on Human Understanding*, , 144.
68. F. J. Cole details Leeuwenhoek's discovery of the parthenogenesis of aphids in "Microscopic Science in Holland in the Seventeenth Century," *Journal of the Quekett Microscopical Club* 4 (1938): 59–77.

69. *New Essays on Human Understanding*, 140–141. See also *Principles of Nature and Grace*, in *Philosophical Essays*, 209.
70. Leibniz to Louis Bourguet, March 22, 1714, in *Die Philosophischen Schriften von Leibniz*, ed. C. I. Gerhardt, 7 vols. (Berlin: Weidmann, 1875–1890), 3:565 as translated by Lloyd Strickland and available through the online resource www.Leibniz-translations.com. On Leibniz's support for ovism see *Reflections on the Doctrine of a Single Universal Spirit* (1702), and for animalculism see *Principles of Nature and Grace* (1715), both in *Philosophical Papers and Letters*, 2:905 and 1037, respectively. Committed to the animalculist position, Leeuwenhoek effectively campaigned for it, writing to Leibniz periodically with new findings. See Ruestow's discussion of these letters from 1715–1716 in "Images and Ideas," esp. 198.
71. The requirement for divine preformation cannot be forgotten amid discussion of Leibniz's dynamic conception of an endless transformation of forms (just as his arguments against metempsychosis cannot), since it is on the basis of such claims that Leibniz consistently rejected vitalism, Van Helmont's "archeus," and epigenesis. See especially his "Considerations on Vital Principles and Plastic Natures, by the Author of the System of Pre-Established Harmony" (1705), in *Philosophical Papers and Letters*, 2:953. For some discussion of Leibniz's earliest thoughts on preformation, see Richard Arthur, "Animal Generation and Substance in Sennert and Leibniz" in *The Problem of Animal Generation in Early Modern Philosophy*, ed. Justin Smith (Cambridge: Cambridge University Press, 2006), 147–174; and Smith, *Divine Machines*, 165–196.
72. "Letter to Hansch on Platonic Philosophy," July 25, 1707, in *Philosophical Papers and Letters*, 2:963. The original here is "Sunt tamen in nobis semina eorum, quae discimus, ideae nempe, & quae inde nascuntur aeternae veritas." Initially included in *Epistola Godefridi Guilielmi Leibnitii ad Michaelem Gottlieb Hanschium*, ed. Georg Veesenmeyer (Leipzig: Gleditsch, 1716), this is more easily found in the 1768 edition of Leibniz's works *Gothofredi Guillelmi Leibnitii, Opera Omnia*, ed. Louis Dutens, vol. 2 (Geneva: Fratre de Tournes, 1768), 223; Loemker has broken one Latin sentence into two in his edition.
73. *New Essays on Human Understanding*, 80. Elsewhere Leibniz will distinguish a "virgin" idea from one that has mingled with others and thereupon generated impossible or superfluous notions. See ibid., 264.
74. For references to Leeuwenhoek's letters surrounding his discovery of parthenogenesis see Ruestow, "Images and Ideas," 219–220nn166–168; for the precise manner in which Leeuwenhoek attempted to use this as a model for the spermatic animalcules, see F. J. Cole, "Microscopic Science in Holland," 64. Charles Bonnet is typically credited with this discovery, for while Leeuwenhoek did in fact first observe that the young were present as miniature adults within the parent—a discovery taken to confirm pre-

NOTES TO CHAPTER 2

existence theory in the 1670s—it was Bonnet, following Réaumur's efforts, who starting in 1740 successfully raised ten generations of aphids without a single male present—that is, parthenogenetically. This led Bonnet to the ovist view of preexistence theory, a position he would modify in favor of his germ theory by 1764.

75. *New Essays on Human Understanding*, 80.
76. Letter to Hansch on Platonic Philosophy, July 25, 1707, in *Philosophical Papers and Letters*, 2:964; on Leibniz's support for Plato's doctrine of recollection see also *Discourse on Metaphysics*, ibid., 1:493. On the relationship between Plato and Leibniz in terms of the ideas and necessary truths, see especially Paul Schrecker, "Leibniz and the Timaeus," *Review of Metaphysics* 4 (1951): 495–505. Leibniz's appreciation for Plato is declared on numerous occasions, indeed the opening passages of his preface to the *New Essays* identifies Aristotle and Locke for holding a position in stark contrast to Plato and Leibniz's shared approach to the mind and its ideas.
77. Thomas Hankins includes an engraving of 'sGravesande's apparatus built to test Leibniz's theory of active force; 'sGravesande's experiments were performed in 1721. See Hankins, *Science and the Enlightenment* (Cambridge: Cambridge University Press, 1985), 31–33.
78. Leibniz argued that force understood as a mere measurement of motion failed to account for the active force or *"vis viva"* describing the accumulated and transferable energy of moving bodies; indeed it was transferability without loss that was necessary for the proper understanding of the conservation of force as a whole. See especially *A Brief Demonstration of a Notable Error of Descartes and Others Concerning a Natural Law* (1686) and *Specimen Dynamicum* (1695), both in *Philosophical Papers and Letters*, 1:455–463 and 2:711–738, respectively. Two helpful discussions of Leibniz's position here are Daniel Garber's "Leibniz: Physics and Philosophy," in *The Cambridge Companion to Leibniz* (Cambridge: Cambridge University Press, 1995), 270–352; and Donald Rutherford, *Leibniz and the Rational Order of Knowledge* (Cambridge: Cambridge University Press, 1995), 237–264.
79. See his "Against Barbaric Physics: Toward a Philosophy of What there actually Is and Against the Revival of the Qualities of the Scholastics and Chimerical Intelligences," in *Philosophical Essays*, 312–320, and the fifth reply to Clarke, numbers 113–116, in *Philosophical Papers and Letters*, 2: 1165–1166.

CHAPTER TWO

80. Significant excerpts of Haller's account of forces are collected by Shirley Roe in *The Natural Philosophy of Albrecht von Haller* (New York: Arno, 1981). Roe reconstructs Haller's position on the forces at work in "animal mechanics" in *Matter, Life, and Generation: Eighteenth-Century Embryology and*

the Haller-Wolff Debate (Cambridge: Cambridge University Press, 1981), 96–102.

81. Louis Bourguet, *Lettres philosophiques sur la formation des sels et des crystaux et sur la génération et le méchanisme organique des plantes et des animaux* (Amsterdam: F. L'Honoré, 1729), esp. 56–168; for "organic mechanics" see p. 64. Medieval philosophers described the work that Aristotle had attributed to the "nutritive soul" as a process of absorption, which they termed "intussusception." This term was later taken up by René Réaumur in 1709 to describe the processes of shell formation in "De la formation et de l'acroissement des coquilles des animaux tant terrestes qu'aquatiques, soit de mer soit de rivière," *Mémoires de la Académie Royale des Sciences*, 1709: 364–400, esp. 366, 370. Bourguet took the term from Réaumur but insisted on the interiority of intussusception (71) in contrast to the kind of external, mechanical accretion occurring in crystals or shell formation. Buffon used the term "intussusception" in line with Bourguet's account of an internal absorption or assimilation (e.g., *History of Animals*, chap. 3, "Of Nutrition and Growth"), as did Kant when arguing that systems may "grow from within (*per intussusceptionem*), but not by external addition (*per appositionem*)" (A833/B861). The appearance of "intussusception" after Kant shows its meaning to have changed again, in this case via Schelling, who used it in his philosophy of nature to identify the universal tendency of attraction in nature. See *First Outline of a System of the Philosophy of Nature* (1799), trans. K. Peterson (Albany: SUNY Press, 2004), 7. A brief review of Bourguet's position is in J. Roger, *The Life Sciences in Eighteenth-Century French Thought*, trans. Robert Ellrich (Stanford, CA: Stanford University Press, 1997), 300–303. For a fuller treatment see François Duchesneau, "Louis Bourguet et le modèle des corps organiques," in *Antonio Vallisneri: L'édizione del testo scientifico d'età moderna*, ed. M. T. Monti (Florence: Leo Olschki, 2003), 3–31. Thomas Hankins describes Buffon's "popularization" of Bourguet's main tenets in *Science and the Enlightenment* (Cambridge: Cambridge University Press, 2005), 128–129.

82. Stephen Hales, *Vegetable Staticks* (London: Scientific Book Guild, 1961). A comprehensive discussion of Hales is in D. G. C. Allan and R. E. Schofield, *Stephen Hales, Scientist and Philanthropist* (London: Scolar, 1980). Henry Guerlac has two excellent essays on Hales in his *Essays and Papers in the History of Modern Science* (Baltimore: Johns Hopkins University Press, 1977), chaps. 12 and 17. Much of this material is reproduced in Guerlac's entry "Hales, Stephen," in the *Dictionary of Scientific Biography*, ed. C. Gillispie (New York: Charles Scribner's Sons, 1970–1980): 35–48.

83. Hales, *Vegetable Staticks*, 192.

84. Ibid., 185–186. Hales thought that temperature provided the mechanical means for respiration: "But as plants have not a dilating and contracting *Thorax*, their inspirations and expirations will not be so frequent as those

NOTES TO CHAPTER 2

of Animals, but depend wholly on the alternative changes from hot to cold, for inspiration, and *vice versa* for expiration." Ibid., 186.

85. A helpful description of Hales's discussion of this is given by Julius von Sachs in his *History of Botany*, 476–482. Von Sachs takes a dim view of the manner in which Hales's contributions to the investigation of air were ultimately understood. In his view Hales's "successors did not comprehend the fundamental importance of these considerations, and made no use of the pregnant idea, that a much larger part of the substance of plants comes from the air and not from the water or the soil. . . . [T]hey quoted and repeated Hales's experiments and observations again and again, but forgot that which in his mind bound all the separate facts together" (481). Hales's experiments were critical in providing the background for Lavoisier's establishment of the oxygen theory of combustion; this connection is established by Guerlac in *Lavoisier—The Crucial Year: The Background and Origin of His First Experiments on Combustion in 1772* (Ithaca, NY: Cornell University Press, 1961).

86. Stephen Hales, *Vegetable Staticks*, 179–180. Boyle had published the results of his experiments with a vacuum pump in 1660, experiments demonstrating the necessity of "elastic" or "good" air for sustaining both a flame and animal life. Since then it had become common to consider air as containing life-sustaining properties, but Hales was original in suggesting it might have a role in generation.

87. Ibid., 203. Hales had earlier suggested the importance of light: "May not light also, by freely entering the expanded surfaces of leaves and flowers, contribute much to the ennobling the principles of vegetables? for Sir *Isaac Newton* puts it as a very possible query, 'Are not gross bodies and light convertible into one another? . . . The change of bodies into light, and of light into bodies, is very conformable to the course of nature, which seems delighted with transmutations. *Opt. qu. 30.*'" Ibid., 186–187.

88. Ibid., 202.

89. Patrick Blair, *Botanick Essays in Two Parts* (London: W. and J. Innys, 1720).

90. Patrick Blair, "Observations Upon the Generation of Plants, in a Letter to Hans Sloan," *Philosophical Transactions* 31 (1720–1721): 216–221. Leeuwenhoek's was no small presence in the Royal Society, and to get a sense of this, one need only count the number of letters he wrote regarding his discoveries that were published in the *Philosophical Transactions* over the years: starting with 1693, 3 letters; 1698, 1; 1700, 5; 1702, 1; 1704, 8; 1706, 5; 1708, 10; 1710, 4; 1712, 1; 1720, 4; 1722, 5. Blair's 1720 piece devoted numerous pages to refuting Leeuwenhoek's position and took up many of the cases Leeuwenhoek himself had described in his published letters. See *Botanick Essays in Two Parts*, 309ff.

91. Blair, "Observations Upon the Generation of Plants," 218. Blair went to great lengths when making this point in 1720: "The features, the gestures, the humours, the tracts of face, the temper, the stature, the voice, the ex-

ternal shape and figure of the body; the inward passions of the mind, the distempers, and frequently the virtuous and vicious inclinations, are as much imparted to us by our mothers as by our fathers; and this is obvious to us every day, in those they call mongrel animals." *Botanick Essays in Two Parts*, 309–310.

92. Camerarius's letter was addressed to Michael Valentin under the title *Academiae Caesareo Leopold. N.C. Hecotorus II. Rudolphi Jacobi Camerarii, Professoris Tubingensis, ad Thessalum, D. Mich. Bernardum Valentini Professorum Giessensem excellentissimum, de sexu plantarum epistola* (Tubingae: Typis Viduae Rommeii, 1694). Excerpts from the letter were published by Valentin in 1696, and the letter was then published unabridged in 1700 and 1701. (This is the same Camerarius mentioned by Leibniz in his letter to Bourguet in 1714; see n. 70.) Today Camerarius's letter is easily found in translation under the title *Ueber das Geschlecht der Pflanzen (De sexu plantarium epistola)*, trans. and ed. M. Möbius (Leipzig: W. Engelmann, 1899). The translator's introduction provides bibliographical information for all editions subsequent to the first appearance of the letter in 1694.

93. Camerarius's impact is traced by A. G. Morton in *History of Botanical Science*, 214–220, 239–245. Morton takes the shared notion of sexuality in plants and animals to mark the beginning of a rapprochement between botany and zoology towards something that will become biology (238).

94. Buffon translated the second edition of the *Vegetable Staticks* (1727, 1731) in 1735, and his French translation was later translated into German under the direction of Christian Wolff in 1748; Wolff's preface, which preceded Buffon's in the German edition, will be discussed later in connection to Kant. In the years following, there would be translations available in Dutch (1750) and Italian (1765). Regarding the widespread availability of Hales, Allan notes, "The editors of the 1809 *Philosophical Transactions, Abridged* observed that an abstract of Desaguilier's review of the *Vegetable Staticks* was unnecessary, 'the work itself being in the library of every person who possesses the least taste for physiological inquiries.'" See Allan and Schofield, *Stephen Hales*, 128.

95. Buffon's changes to Hales's text are discussed by Allan in ibid., 130–131. The best general resource for Buffon remains Jacques Roger's *Buffon: A Life in Natural History*, trans. Sarah Bonnefoi (Ithaca, NY: Cornell University Press, 1997).

96. Georges Buffon, "Préface du traducteur," in *La Statique des végétaux, et l'analyse de l'air*, by Stephen Hales (Paris: J. Vincent, 1735), 3. A complete English translation of Buffon's preface with editorial notes is available in Phillip Sloan and John Lyon, *From Natural History to the History of Nature: Readings from Buffon and His Critics* (Notre Dame, IN: University of Notre Dame Press, 1981), 35–40.

97. Georges Buffon, preface to *La methode des fluxions, et des suites infinies*, by Isaac Newton (Paris: Debure l'aîné, 1740), esp. xv–xxv. Buffon's opening

NOTES TO CHAPTER 2

discussion concerns his decision in dating Newton's piece to sometime between 1664 and 1671, followed by remarks on the special problem of infinity in mathematics. Buffon's case for Newton as the true author of the calculus details numerous letter exchanges and other pieces of evidence in Newton's favor.

98. Georges Buffon, "Initial Discourse" (1749), in *Histoire naturelle, générale et particulière* (Paris: Imprimerie Royale, 1749–1767), vol. 1. (Buffon's enormously popular work eventually totaled thirty-six quarto volumes with supplements, 1749–1788; by the end of the eighteenth century it had been translated into German, English, Spanish, Italian, and Dutch.) Although the *Natural History* was fairly quickly translated into well-known English editions—William Smellie's in Edinburgh and J. S. Barr's in London—the "Initial Discourse" was not included in any of the many English editions it went through. There is a complete English translation of the "Initial Discourse" with editorial notes in Sloan and Lyon, *From Natural History to the History of Nature*, 89–128, and separately as "The 'Initial Discourse' to Buffon's *Histoire Naturelle*: The First Complete English Translation," by John Lyon, in *Journal of the History of Biology* 9 (1976): 133–181.
99. Georges Buffon, "Initial Discourse," in Sloan and Lyon, *From Natural History to the History of Nature*, 125.
100. Ibid., 121–122.
101. Ibid., 123. It is tempting to read Hume's influence here even with the significant difference that for Hume such probability could only yield "belief." According to Roger, at least, Buffon was not yet aware of Hume's work in either the *Treatise* (1739–1740) or the *Enquiry* (1748). See Roger's *Buffon*, 90.
102. Buffon took "the most beautiful and felicitous" use of the true method to be Newton's theory of gravity: "We must admit that if Newton had only given us the physical conformations of his system without having supported them by precise mathematical evaluations they would not have had nearly the same force." "Initial Discourse," in Sloan and Lyon, *From Natural History to the History of Nature*, 125.
103. As Buffon expressed it, "The penetrating forces by which these immense bodies are animated, by which they act reciprocally upon each other at a distance, animate at the same time every particle of matter; and this mutual propensity of all parts toward each other, is the first bond of beings, the principle of consistency and permanency in nature, and the support of harmony in the universe." "Second View of Nature" (1765), in *Natural History, General and Particular*, trans. William Smellie (Edinburgh: W. Creech, 1780), 7:93.
104. Georges Buffon, "The Ox" (1753), in ibid., 3:423–424.
105. Bourguet had insisted that something like an internal mold (*Moule*) determined the organization of the organic material when forming the organism and that this kind of mold was unique to organic mechanism. Whereas

a crystal simply repeated the same mold or shape over and over again, the organism needed a different means for accounting for the innumerable different parts of its organization. This is not to say that the organism was reducible to its organization for Bourguet; as a Leibnizian, Bourguet took the entire process to be a case of mechanical accommodation to an underlying dominant monad. See Bourguet, *Lettres philosophiques*, 146, 165.

106. Georges Buffon, "Of Nutrition and Growth," chap. 3 of *History of Animals*, (1749), in *Buffon's Natural History*, trans. J. S. Barr (London: H. D. Symonds, 1797), 2:302.

107. Thomas S. Hall traces the eighteenth-century use of such "inexplicable explicative devices," focusing, for example, on Haller's appeal to "irritability" as an explanation for muscle contraction. See "Biological Analogs of Newtonian Paradigms," *Philosophy of Science* 35 (1968): 6–27. For specific discussion of Buffon's application of Newton's method, see Gasking, *Investigations into Generation*, 92–94.

108. Buffon, "Of Nutrition and Growth," 303. Buffon repeatedly defended the use of analogies in line with Newton's own practice of reasoning. For example, "In my theory of expansion and reproduction, I first admit the mechanical principles, then the penetrating force of gravity, which we are obliged to accept, and, from analogy and experience, I have concluded the existence of other penetrating forces peculiar to organized bodies." "Of Nutrition and Growth" (1749), chap. 3 of *History of Animals*, in *Natural History, General and Particular*, trans. Smellie, 2:48.

109. Georges Buffon, "A Comparison Between Animals, Vegetables, and Other Productions of Nature," chap. 1 of *History of Animals*, in *Buffon's Natural History*, trans. Barr, 2:272.

110. Georges Buffon, "Of Reproduction in General," ibid., 2:279.

111. This caused some to associate Buffon's theory with the *"homoeomeries"* ("things with like parts") attributed to Anaxagoras. Focusing on Buffon's attention to crystal formation, Haller, for example, aligned Buffon's account with Anaxagoras: "The *Homeomeria* of Anaxagoras evidently govern this portion of nature, wherein one sees particles form into a whole with constancy and regularity, without the least suspicion of a semen or seed being involved." "Reflections on the Theory of Generation of Mr. Buffon," trans. Phillip R. Sloan, in *From Natural History to the History of Nature*, 315. Roger similarly identifies Buffon's organic molecules and the homoeomeries of Anaxagoras in *Life Sciences in Eighteenth-Century French Thought*, 667n92. It seems likely, however, that it is rather Bourguet who is once more serving as an influence for Buffon's account. See Bourguet, *Lettres philosophiques*, 56–57. Anaxagoras in fact conceived of seeds (*"homoeomeries"* was the term used to describe Anaxagoras's seeds by Aristotle, among others) as necessarily containing both the substance from which they came and a "portion" of their opposites, a necessity if change was to be explained (e.g., food transformed into hair and flesh). See the selec-

tions and commentary in "Anaxagoras of Clazomenae," in *The Presocratic Philosophers*, 2nd ed., trans. and ed. G. S. Kirk, J. E. Raven, and M. Schofield, (Cambridge: Cambridge University Press, 1983), 368–371, 374–378. It was this inclusion of opposites that presumably led to Goethe's interest in Anaxagoras's account. See "On the Spiral Tendency in Plants," in *Goethe's Botanical Writings*, trans. Bertha Mueller (Woodbridge, CT: Oxbow, 1952), 131.

112. The lack of entelechy here would be one of the main avenues for Haller's critique of Buffon, one launched first in an anonymous French review and then added as a preface to the German translation of Buffon's *Natural History* volumes overseen by Abraham Kästner (1750–1774). Rehearsing Buffon's discussion of the internal mold and the penetrating force, Haller complained that these could not provide a reasonable source of organization given the complexity of the body. "In brief," Haller concluded, "what is the cause which arranges the human body in such a way that an eye is never attached to the knee, an ear is never connected to the hand, a toe never wanders to the neck, or a finger is never placed on the extremity of the foot, as happens in the crystallization of salts, where one finds at all times some of the spires sometimes similar and sometimes different, often without form, and in reverse order?" "Reflections on the Theory of Generation," 320.

113. The fact that hydra—a name chosen by Linnaeus for the Medusa-like properties of the polyp—both reproduced and regenerated by budding had been first described (with little notice taken) by Leeuwenhoek in a 1702 letter to the Royal Society. See "Part of a Letter from Mr. Antony van Leeuwenhoek, F. R. S. Concerning Green Weeds Growing in Water, and Some Animalcula Found about Them," *Philosophical Transactions* 23 (1702–1703): 1304–1311. Once published, Trembley's own investigations, by contrast, had an enormous and immediate impact on the intellectual life of the eighteenth century. See *Mémoires pour servir a l'histoire d'un genre de polypes d'eau douce* (Paris, 1744).

114. Unlike Buffon, preexistence theorists took the entire individual to be formed in either the male or female parent and thus suffered to explain the appearance of inheritance from both sides. See the introduction to the present volume, n. 3, on the difference between preexistence and preformation, distinctions frequently conflated within the secondary literature. The fact that Buffon's organic molecules were themselves preformed or premolded in the parents might have contributed to Barr's decision to usually translate "développer" as "unfold"—a verb choice typically used alongside "evolution" and "development" when referring to the expansion of a microscopic yet fully formed *preexistent* individual.

115. Haller, "Reflections on the Theory of Generation," 322.

116. Although they disagreed on key points regarding generation—Maupertuis claimed never to have understood Buffon's notion of internal molds—the

influence was mutual during these years. This is made especially clear in John Turbeville Needham's descriptions of his work with Buffon, "A Summary of some late Observations upon the Generation, Composition, and Decomposition of Animal and Vegetable Substances," *Philosophical Transactions of the Royal Society* 45 (1748): 615–666, esp. §18, p. 633: "He [Buffon] had long been disatisfy'd with the Opinion of pre-existent Germs in Nature; and he and Mr. *Maupertuis*, President of the *Academy* of *Sciences* at *Berlin*, had often discours'd together upon the Subject. We have several Hints of this Dissatisfaction, in a little Book, published by Mr. *Maupertuis* himself upon this Question at *Paris*, before my Arrival there; in short, it was by general Reflections, and some other consequent Thoughts, that Mr. *de Buffon* was conducted to frame his System of organical Parts." Thus Buffon, for example, praised the arguments against preexistence theories laid out in Maupertuis's *Venus Physique*, declaring that "this author is the first who has returned into the road of truth, from which we were farther strayed than ever, since the supposition of the egg system, and the discovery of spermatic animals." "Experiments on the Method of Generation," chap. 6 of *History of Animals* (1749) in *Buffon's Natural History*, trans. Barr, 3:76. Once the *Natural History* was published, Buffon was cited by Maupertuis in support of his argument for some fluidity between the animal and vegetable classes. See Maupertuis's *Essai sur la formation des corps organisés*, in *Oeuvres* (Paris, 1756), vol. 2, §47, p. 152, where he refers to Buffon's vol. 2, chap. 8, p. 303, and chap. 9, p. 322 (corresponding to Barr's translation, *Buffon's Natural History*, 3:206–207 and 223–224, respectively). Note, however, Buffon's explicit rejection of appeals to such fluidity on the basis of the Tree of Diana, in "Initial Discourse," in Sloan and Lyon, *From Natural History to the History of Nature*, 101.

117. Mary Terrall describes the difficult social and political environment within which Maupertuis worked in *The Man Who Flattened the Earth: Maupertuis and the Sciences in the Enlightenment* (Chicago: University of Chicago Press, 2002), esp. chaps. 9 and 10. A comprehensive listing of Maupertuis's unusually complicated publication history—a history comprising multiple editions under different titles, often published anonymously or even pseudonymously—is in Giorgio Tonelli's introduction to *Oeuvres*, by P. L. Moreau de Maupertuis (Hildesheim: G. Olms, 1974), 1:xi–lxxxiii. For our purposes it is enough to trace the trajectory of Maupertuis's publications regarding generation theory. These started in 1744 with three successive pamphlet editions of an anonymously written discussion of biological generation, a discussion occasioned by the sensation created in Parisian salon culture by an albino boy (born to African slaves living in colonial South America) who had been paraded around Paris by the aristocracy as a curiosity that year. See "Dissertation physique à l'occasion du nègre blanc" (Leiden, 1744). The following year this piece was reissued (and slightly changed) as "Concerning the Origin of Animals" and printed to-

gether with a separate essay on "Varieties in the Species of Man" under the title *Venus Physique*; this too was published anonymously, with neither publisher nor location identified, only the date, 1745. In 1751, Maupertuis returned to the issue, this time publishing his essay in Latin as a (fake) thesis from Erlangen written by a student identified only as Dr. Baumann; the original is no longer extant, but at the time, it was published (in Berlin, not Erlangen) as *Dissertatio inauguralis metaphysica de universali naturae systemate* (Berlin, 1751). An edition of Maupertuis's collected works that appeared in 1752 included the *Venus Physique*, thus revealing Maupertuis's authorship. See *Oeuvres* (Dresden: C. Walther, 1752). In 1754, Maupertuis anonymously reissued the Baumann thesis from 1751, now under the title *Essai sur la formation des corps organisés* (Berlin, 1754). The *Essai* was reissued two years later under the title *Systêm de la nature* for the next edition of Maupertuis's collected works, revealing thereby Maupertuis's authorship of the *Essai* (and the Baumann thesis). See *Oeuvres*, 4 vols. (Lyon, 1756). Finally, in 1761, an anonymous translator working out of Potsdam issued a German translation of the Baumann thesis as *Versuch von der Bildung der Körper, aus den Lateinischen des Herrn von Maupertuis übersetzt von einem Freunde der Naturlehre* (Leipzig, 1761); it was this edition that Kant owned.

118. Pierre Maupertuis, *The Earthly Venus*, trans. S. Boas (New York: Johnson Reprint, 1966), 56. Like Buffon, Maupertuis agreed that female particles must contribute to the formation of an embryo, and the *Venus Physique* attacked preexistence doctrines for their inability to account for obvious cases of "blended" offspring. In his later discussions Maupertuis included extensive documentation of the Ruhe family in Berlin, since the many cases of polydactylism in the family were shown to be heritable on both maternal and paternal lines, evidence Maupertuis took to be fatal for both spermist and ovist versions of the preexistence theory of generation. See *Essai sur la formation des corps organizes*, 159–161. New data regarding the Ruhe family was included only in the 1754 *Essai sur la formation des corps organisés* Berlin edition, as evidenced by its absence from the German translation of the Baumann thesis. On the Ruhe family see, for comparison, Maupertuis, *Versuch von der Bildung der Körper*, §37, p. 31. Section numbers in the 1761 German translation correspond to the 1754 Berlin edition until §38, after which sections are added in the remainder of the French edition with the result that German §39 = French §40, German §53 = French §55, German §62 = French §65.

119. Réaumur's fame as a naturalist was due to the publication of his multivolume work on insects, *L'histoire des insectes* (Paris, 1734), at which time he separately published a treatise on incubation. The quote is from his discussion of philosophical accounts of generation that closed the treatise on incubation, "Quatriéme mémoire: Esquisse des amusemens philosophiques que les oiseaux d'une basse-cour ont à offrir," in *Art de faire éclorre et d'élever*

en toute saison des oiseaux domestiques de toutes especes (Paris: Imprimerie Royale, 1749), 2:329–330. Abraham Trembley prepared an English translation of this work: *The Art of Hatching and Bringing Up Domestic Fowls by Means of Artificial Heat* (London, 1750), 461–462.
120. Réaumur, "Quatriéme mémoire," 330; Réaumur, *The Art of Hatching*, 462.
121. In the preface to the German edition of the Baumann thesis, for example, the translator declared the two positions to be identical—"Die hauptsache scheint mir der Monadologie des Herrn v. Leibnitz einerley zu seyn"—even as Maupertuis was praised for his added doctrines. Maupertuis, *Versuch von der Bildung der Körper*, paragraph 5 (preface unpaginated). Interest in Leibniz's monads was widespread as a result of 1747's prize question "for or against the doctrine of monads," a contest overseen by Maupertuis himself as president of the Berlin Academy of Sciences. As Euler, the academy's leading mathematician, described it, "There was a time when the dispute about monads was so lively and general that one spoke of them heatedly in all social circles, even in the *corps de garde*. There was almost not a single lady at court who had not declared herself for or against monads." Quoted in Terrall, *Man Who Flattened the Earth*, 258.
122. Maupertuis, *Versuch von der Bildung der Körper*, §§3, 4, p. 15. Later Maupertuis appears to be responding directly to Réaumur, explicitly listing the impossibility of understanding the eye or the ear to be the result of such forces. Ibid., §14, p. 21.
123. Ibid., §33, pp. 30–31.
124. For discussion of Leibniz and Maupertuis, see Terrall, *Man Who Flattened the Earth*, 328–334. I agree with Terrall's assessment regarding the actual nature of Maupertuis's understanding of the particles' organic forces as a conception that does not, in the end, rely on the kind of metaphysics associated with Leibniz's panorganicism. As Terrall puts it, "When mechanical properties proved insufficient to the explanatory task at hand, Maupertuis simply attached various appropriate active properties directly to matter, for a different kind of active matter" (333; see also her comments at n. 77). It is perhaps not technically correct, therefore, to identify Maupertuis's position with hylozoism, for despite the divinely (and hierarchically) granted intelligence ascribed to the particles, their "life" is in the end only an activity that has been somehow "attached" to matter. By contrast, John Zammito takes Maupertuis's commitment to hylozoism to be both real and the basis of Kant's later rejection of Maupertuis's views on generation. "Kant's Early Views on Epigenesis: The Role of Maupertuis," in *The Problem of Animal Generation in Early Modern Philosophy*, ed. Justin E. H. Smith (Cambridge: Cambridge University Press, 2006), 317–354.
125. G.-L. Buffon, "Of the Expansion, Growth, and Delivery of the Foetus," chap. 11 of *History of Animals*, in *Natural History, Genetic and Particular*, trans. Smellie, 2:308.
126. Ibid., 305.

127. Ibid., 309. While Leibniz's explicit discussion of "analysis situs" is readily available today (e.g., *Philosophical Papers and Letters*, 1:390; see also 1:382), it is hard to say whether anyone actually read it in the eighteenth century. Louis Dutens's well-regarded edition of Leibniz's complete works appeared only in 1768 and included neither Leibniz's essay nor his letter to Huygens from the same period, a letter that also made mention of a new geometry of situation (Leibniz to Huygens, September 8, 1679, in *Philosophical Papers and Letters*, 1:381–390). Vincenzo De Risi traces the impact of Leibniz's rumored geometry in *Geometry and Monadology: Leibniz's Analysis Situs and Philosophy of Space* (Basel: Birkhäuser, 2007). According to De Risi, "It was thanks to Wolff that the entire scientific world first learned about *analysis situs* and attached it to Leibniz's fame," for in volume 1 of Wolff's *Elementa matheseos universae* (Halle: 1713–1715), Wolff had described Leibniz's "new analysis of situation, constructed upon a peculiar kind of calculus (which he calls calculus of situation), completely different from the calculus of magnitudes." See De Risi, *Geometry and Monadology*, 94–95; see also n. 104.
128. A brief, clear discussion of "the general logic of embryo geography," including topographical mapping techniques using applied geometry, is in Sean B. Carroll's *Endless Forms Most Beautiful: The New Science of Evo Devo and the Making of the Animal Kingdom* (New York: W. W. Norton, 2005), 89–98.
129. Buffon, "Second View of Nature," 7:97.
130. Ibid., 7:101.
131. The best discussion of Buffon's relationship to Linnaeus is Phillip R. Sloan's "The Buffon-Linnaeus Controversy," *Isis* 67 (1976): 356–375.
132. Buffon, "Initial Discourse," in Sloan and Lyon, *From Natural History to the History of Nature*, 104.
133. Ibid., 105. Between the inability to determine criteria capable of determining essential divisions between species (since this required agreement regarding their "essence") and the empirical experience of the fluidity of forms (an experience frequently undermining belief in fixed essences at all), deep tensions within taxonomy had arisen by midcentury between arbitrarily determined criteria like reproductive organs and conflicting experience with respect to claims regarding biological affinity. The "artificial" system of Linnaeus self-consciously took this all for granted, but even though Linnaeus grasped both the logical problem and the practical tensions within taxonomical science, the difficulties were in fact exacerbated by his defaulting, as a matter of practical necessity, to the idea of species fixity. As a result, as Sachs puts it, species fixity "became with his [Linnaeus's] successors an article of faith, a dogma, which no botanist could even doubt without losing his scientific reputation; and thus during more than a hundred years the belief that every organic form owes its existence to a separate act of creation and is therefore absolutely distinct from all other forms, subsisted side by side with the fact of experience, that there is an intimate tie of relationship between these forms which can only be im-

perfectly indicated by definite marks. Every systematist knew that this relationship was something more than mere resemblance perceivable by the senses, while thinking men saw the contradiction between the assumption of an absolute difference of origin in species (for that is what is meant by their constancy) and the fact of their affinity." Sachs, *History of Botany*, 10. Morton details Linnaeus's awareness of this problem, describing its resolution as the precondition for a shift toward the concept of organic evolution. See *History of Botanical Science*, 262–276, esp. 270.

134. Buffon, "Initial Discourse," in Sloan and Lyon, *From Natural History to the History of Nature*, 102. Cf. "Nature has neither classes nor species; it contains only individuals. These species and classes are nothing but ideas which we have ourselves formed and established." Buffon, "Of Infancy," chap. 2 of *History of Man*, in *Buffon's Natural History*, trans. Barr, 3:326.

135. Buffon, "Initial Discourse," in Sloan and Lyon, *From Natural History to the History of Nature*, 111.

136. Buffon, "Second View of Nature," 96. Following Buffon's distinction between an eternal and a temporal view of a species depends as much on context as anything else. On Buffon's failure to make the difference explicit, see the helpful discussion in Roger's *Buffon*, 325–329.

137. Buffon, "Of the Degeneration of Animals" (1766), in *Natural History, General and Particular*, trans. Smellie, 7:395.

138. Ibid., 398, 400.

139. Ibid., 416, cf. 420, 422.

140. Ibid., 422.

141. Ibid., 415.

142. Ibid., 437.

CHAPTER THREE

143. See Manfred Kuehn, *Kant: A Biography* (Cambridge: Cambridge University Press, 2000), 179.

144. Johann Herder, *Herders Sämmtliche Werke*, ed. B. Suphan (Berlin: Weidmann, 1893), 18:325.

145. Kuehn, *Kant*, 179.

146. *Discours sur la différente figure des astres avec une exposition des systems de MM. Descartes et Newton* (Paris, 1732; 2nd ed., 1742). Kant cited a passage from this text (1:232) that he had found in a review in *Nova Acta Eruditorum* of the 1744 edition (*Ouvrages Divers*, 1 vol. (Amsterdam, 1744)) of Maupertuis's works. See *Nova Acta Eruditorum*, 1745, 221–229. For the original passage see Maupertuis, *Oeuvres* (Lyon: Bruyset, 1756), 1:88. Kant might already have been familiar with the first version of this essay, since it was included in the first publication (1746) of the Berlin Academy of Sciences under Maupertuis's direction; the annual volume was a central source of income for the Berlin academy and was therefore distributed widely to

NOTES TO CHAPTER 3

booksellers and libraries. See Maupertuis, *Essay on Cosmology*, in *Histoire de l'Académie Royale des Sciences et Belles-Lettres* (Berlin, 1746), 267–294.

147. See *The Only Possible Argument in Support of a Demonstration of the Existence of God* (2:113–116) and *Dreams of a Spirit-Seer Elucidated by Dreams of Metaphysics* (2:327–333).

148. A lengthy discussion of Kant's early views on forces is in Martin Schönfeld's *The Philosophy of the Young Kant: The Precritical Project* (Oxford: Oxford University Press, 2000); Manfred Kuehn looks at Kant's 1747 essay in connection with Kant's teacher Martin Knutzen in "Kant's Teachers in the Exact Sciences," in *Kant and the Sciences*, ed. Eric Watkins (Oxford: Oxford University Press, 2001), 11–30.

149. The so-called *Spin-Cycle* essay's complete title runs, "Investigation of the question whether the earth in its axial rotation, whereby it causes the change of day and night, has experienced any change since the earliest times of its origin, and how one could answer this question, announced for the current year's prize, by the Royal Academy of Sciences in Berlin" (1:183–192).

150. See Mary Terrall's detailed account of Maupertuis's difficulties in Berlin in *Man Who Flattened the Earth*, 231–269. A list of the academy's prize essay questions—and the winners—during Maupertuis's tenure as president, alongside a thorough history of the academy itself, is in Adolf Hartnack's *Geschichte der Königlich Preussischen Akademie der Wissenschaften zu Berlin*, 4 vols. in 3 (Berlin: Reichsdruckerei, 1900); prize essays listed in vol. 1, 409ff.

151. "Les loix du mouvement du repos, deduites d'un principe métaphysique, par M. de Maupertuis, (1746), in *Histoire de L'Académie Royale des Sciences et des Belles-Lettres* (Berlin: chez Ambroise Hande, 1748), 267–294.

152. See "Accord de différentes loix de la nature qui *avoient* jusque'ice paru incompatibles," in *Mémoires de l'Académie Royal des Sciences* (Paris, 1744), 417–426.

153. "Les Loix du Mouvement du repos," 290.

154. Only the first two sections of the older pieces would be republished as the *Essay on Cosmology*. A complete history of this is in Giorgio Tonelli's introduction to Maupertuis's *Oeuvres* (Hildesheim: Georg Olms Verlag, 1974), 1:lix and following.

155. Maupertuis, "Les loix du mouvement et du repos deduites des attributs de la supreme intelligence," *Essay de Cosmology*, in *Oeuvres* (Dresden: George Conrad Walther, 1752), 22–23.

156. Maupertuis's general strategy regarding proofs for the existence of God would be repeated again by Kant in the *Only Possible Argument* (1763).

157. Kant refers to Maupertuis repeatedly in the precritical years, most often invoking Maupertuis's work on celestial mechanics. See *Universal Natural History*, 1:232, 236, 254, 255; *Only Possible Argument*, 2:98, 115, 141; *Dreams*

NOTES TO CHAPTER 3

of a Spirit-Seer, 2:330; *Of the Different Races of Human Beings*, 2:431; *Lectures on Logic*, 16:103, 281 (marginalia); *Nachlass*, 18:230, 18:578; and *Reflexionen zur Medizin*, 15:955.
158. *Oeuvres* (1752), 23.
159. *Statick der Gewächse oder angestelte Versuche mit dem Saft in Pflanzen und ihren Wachstum* (Halle: Rengerischen Buchhandlung, 1748). The translation of Hales (from Buffon's French edition) was done under Christian Wolff's supervision. Wolff had himself published on vegetable anatomy in *Allerhand nützliche Versuche* (Halle, 1721) and on plants in general in his *Vernünftige Gedanken von der Wirkungen der Natur* (Magdeburg, 1723). Some discussion of Wolff's botanical writings can be found in Julius von Sachs's *History of Botany*, 472–476.
160. Kant, like the majority of Hales's readers, valued *Vegetable Staticks* for the many experiments described by Hales, experiments whose results lent themselves to ready application. Kant appealed to Hales in *The Question Whether the Earth Is Aging, Considered from the Point of View of Physics*, 1:208; *Universal Natural History*, 1:326; *Succinct Exposition of Some Meditations on Fire*, 1:381, 382; *New Elucidation of the First Principles of Metaphysical Knowledge*, 1:408; *History and Natural Description of the Most Curious Occurrences Associated with the [Lisbon] Earthquake*, 1:457; and *Opus Postumum*, 21:238n. There are also numerous indirect references to Hales's work: to list only a few, 2:107, 21:454, 21:263, 21:499. It is a measure of Hales's significance that Kant held onto his translation of Hales despite the fact that he is known to have sold books during these years to help make ends meet. It was not until 1766 that Kant's fortunes changed for the better. Taking on the job of sublibrarian at the Schlossbibliothek—the palace library doubled as the university's—meant Kant had the means to move into rooms above Kantor's bookshop. Kantor allowed his tenants to read the books and many periodicals coming through the shop, a fact of its own significance when considering Kant's intellectual development, and a point to be underscored when considering the relatively few books Kant's personal library contained. For something of Kant's circumstances during this period, see Kuehn, *Kant*, 159–160. Kant's library has been cataloged by Arthur Warda in *Immanuel Kants Bücher* (Berlin: M. Breslauer, 1922).
161. Buffon's preface to *Statick der Gewächse*, 8 (unpaginated in text).
162. Wolff's preface to *Statick der Gewächse*, 1 (unpaginated in text).
163. Ibid., 4.
164. Ibid., 6.
165. Kant referred to Buffon repeatedly during his career, most often concerning questions related to natural history. He cited Buffon on cosmology, earthquakes, and geology (1:227, 238, 345, 421, 438, 444, 451); on generation, female attraction, and the geometry of *"analysis situs"* (2:4, 8, 115, 142, 237, 377, cf. 2:38, 429); on volcanoes on the moon and the genealogy

NOTES TO CHAPTER 3

of dogs (8:74, 75, 168); on physical geography (9:213, 303); on the formation of mountains (14:547, 589, 607); on anthropology (15:389); regarding the *Conflict of the Faculties* (15:389); and in the Nachlass (17:96).
166. Haller's prefaces are available in English translation. See "Reflections on the Theory of Generation of Mr. Buffon," trans. Phillip R. Sloan, in *From Natural History to the History of Nature*, , 320.
167. Kant explained that he was first inspired to give such a course after reading a 1751 review of William Wright of Derham's cosmological treatises in the *Hamburgischen freien Urtheile* (1:231).
168. In his opening discourse Buffon appraised the cosmological theories of Newton, Burnett, Woodward, Whiston, and Leibniz's *Protagaea*.
169. *Natural History, General and Particular*, trans. Smellie, 1:64. See also the editorial comments regarding Kant's passage, 1:549ff.
170. As a lecturer without an official position at the university, Kant needed to advertise his upcoming courses, advertisements that included short essays: *West Winds* (1757), *New Theory of Motion and Rest* (1758), *Reflections on Optimism* (1759), *M. Immanuel Kant's Announcement of the Programme of his Lectures for the Winter Semester 1765–66* (1765), and *On the Different Races of Mankind* (1775).
171. Kant's longtime interest in cosmology would find its place, albeit negatively, in the first *Critique*'s discussion of the antinomies. For more discussion of Kant's early cosmological theories, see Schönfeld's *Philosophy of the Young Kant*; Alison Laywine, *Kant's Early Metaphysics and the Origins of Critical Philosophy*, North American Kant Society Studies in Philosophy, vol. 3 (Atascadero, CA: Ridgeview, 1993); and Michael Friedman, *Kant and the Exact Sciences* (Cambridge, MA: Harvard University Press, 1992), 1–52.
172. *Versuch von der Bildung der Körper, aus dem Lateinischen, des Herrn von Maupertuis übersetzt von einem Freunde der Naturlehre* (Leipzig, 1761). This was a translation of Maupertuis's pseudonymously published Latin "dissertation" from 1751, the so-called Baumann thesis. In 1754 Maupertuis published the thesis in French under the title *Essai sur la formation des corps organisés*, and the essay was retitled for the publication of Maupertuis's collected works in 1756 as the *Système de la nature* with some additions to the text. The German translation was directly taken from the first Latin thesis—of which, according to the German translator, only ten copies were printed—and thus did not include the empirical studies of sexdigitalism, for example, added by Maupertuis for the French editions. Further information regarding the publication history of these pieces can be found in nn. 117 and 118.
173. A thorough account of the Haller-Wolff debates is in Shirley Roe's *Matter, Life, and Generation*, including a publication history of Wolff's works and their subsequent reviews by Haller.
174. Although Kant's 1755 *Universal Natural History* had been reviewed, few copies made it into the hands of potential readers because the press was

impounded—and Kant's books with it—for bankruptcy just after the copies had been printed. A subsequent fire resolved the publisher's financial difficulties, but it also burnt most of the copies of Kant's book. For more of this history see Kuehn's *Kant*, 104–105.

175. English translations from Kant's precritical period—when they are available—are either taken whole from or are based on the translations by David Walford in *Immanuel Kant: Theoretical Philosophy, 1755–1770* (Cambridge: Cambridge University Press, 1992) or Lewis White Beck in *Kant's Latin Writings: Translations, Commentaries, and Notes* (New York: P. Lang, 1992).

176. The encasement, or *"emboîtement,"* theory was first proposed by Malebranche, and both he and Leibniz after him used the verb *"développer"* to describe the change from microscopic to normally sized individuals (for Leibniz, *développer* was in contrast to the *"envelopper"* of the monads at death). When Buffon described this kind of "augmentation" of individuals according to the encasement theory, he used *"développer,"* but—and this decision would create a challenge for his interpreters—he also reappropriated the verb as more properly understood to describe his *own* position regarding the generation and growth of individuals, processes that, according to Buffon, resulted from the joint efforts of the organic molecules and the molded parts collected from both the mother and father. Kästner's translation of Buffon—the only German translation until 1775—regularly chose *"auswickeln"* (literally "to unwrap or unwind," used in the vernacular with respect to unwrapping a swaddled child but usually translated in English as "to unfold") over *"entwickeln"* ("develop" in English) when translating *"développer."* In the early English editions of Buffon, *développer* received divergent translations; the Smellie translations, for example, used "expansion" for *"développer,"* a choice emphasizing augmentation, whereas the Barr translations relied on "unfold," a choice that, given its history, linked Buffon to preexistence theory for anyone unaware of his reappropriation of the word for his own ends. Thus although Buffon was explicitly critical of encasement theories in either their ovist or animalculist versions, distinctions between Buffon and preexistence theories were easily obscured from the start because of the language in play. On this point compare, for example, Buffon's opening passages in the *History of Animals*, chap. 3, "Of Nutrition and Growth," between editions. See Buffon, *Histoire naturelle* (1749), 2:41, and Kästner's translation (1750), vol. 1, part 2, p. 27; Smellie's translation (*Natural History, General and Particular*, 1785), 2:39; and Barr's translation (*Buffon's Natural History*, 1791), 2:298. Today, English-language historians of science refer to *"emboîtement"* or encasement theory as a theory of "individual preformation," "evolution," "development," and "unfolding," often obscuring, thereby, the particular histories associated with each of these terms.

177. In these passages David Walford has translated *Fortpflanzung* ("reproduction" in English) as "propagation." Within the context of Walford's trans-

lation as a whole, I think that this choice might be misleading at points, although propagation is good for capturing the nonsexual nature of reproduction according to encasement theory. Kästner used *"Vermehrung"* as a translation of Buffon's description of the "augmentation" of an embryo in preexistence theories. The taking in of nutrition, for example, yields "eine Vermehrung" and "diese Vermehrung der Größe nennet man das Auswickeln, weil man sie dadurch zu erklaren hat, daß man sagte, das Tier sey in kleinen gebildet, wie es seiner völligen Größ nach beschaffen ist, und daher, liese sich leicht begreifen, wie sich seine Theile auswickelten, indem nach und nach eine dazu kommende Materie alle in gehörigen Ebenmaaße vergrößerte," in *Allgemeine Historie der Natur: Nach ihren besonderen Theilen abgehandelt*, trans. A. G. Kästner, vol. 1, part 2 (Hamburg: G. C. Grund and A. H. Holle, 1750), 27.

178. A helpful discussion of Kant's attempt to synthesize preexistence theory and epigenesis in this section is in Mark Fisher, "Kant's Explanatory Natural History: Generation and Classification of Organisms in Kant's Natural Philosophy," in *Understanding Purpose: Kant and the Philosophy of Biology*, ed. Philippe Huneman, North American Kant Society Studies in Philosophy, vol. 8 (Rochester, NY: University of Rochester Press, 2007), 101–121.

179. Paul Menzer takes Kant—wrongly, in my view—to have Caspar Wolff's position rather than Hales's in mind in this passage. See Menzer, *Kants Lehre von der Entwicklung in Natur und Geschichte* (Berlin: G. Reimer, 1911), 104. That said, in Herder's notes from Kant's lectures on metaphysics during the same period as the 1763 piece it is clear that, without naming them, Kant could have understood that the specific difficulty facing Haller and Wolff was the lack of any decisive evidence in favor of one position versus the other. As Herder recorded him, *"Die Physikalischen beobachtungen zeigen, daß der Körper zuerst gebildet wurde, andere daß sie bei der Schöpfung gebildet sei"* (29:889). In his notes Herder went on to report that the main conceptual difficulty facing the life sciences was twofold, at least so far as Kant understood their attempt to discern the processes of generation, namely, the conception of freedom on the one hand and its generation in the world (*die Zeugung seines gleichen im Raum*) on the other.

180. In Kant's words, "Große kunst und eine zufällige Vereinbarung durch freie Wahl gewissen Absichten gemäß ist daselbst augenscheinlich und wird zugleich der Grund eines besondern Naturgesetzes, welches zur künstlichen Naturordnung gehört. Der Bau der Plflanzen und Thiere zeigt eine solche Anstalt, wozu die allgemeine und nothwendige Naturegesetze unzulänglich sind" (2:114).

181. Compare Wolff's opening reverie regarding a metaphysical knot (*ein unauflöslicher Knoten*) so tangled that one might be tempted "to simply cut or hack through it": "oder, wenn wir ihn aufzuwickeln nicht vermögend sind, ihn gleich zerschneiden, oder gar zerhauen wollen." Like Kant, Wolff's

knot referred to the difficulty in connecting the supersensible and material realms. By 1766, however, Kant was prepared to dismiss the supposed entanglement as a case of the mind's "surreptitious" application of sensible concepts to intelligible objects (2:321n; cf. 10:72). See Wolff's preface to the translation of Stephen Hales's *Vegetable Staticks, Statick der Gewächse*, 2 (unpaginated in text).

182. Kant taught "anthropology" as an independent course starting in 1772 and continued to teach it every year until his retirement in 1796. A careful reconstruction of Kant's developing interests in anthropology during these years is in John Zammito's *Kant, Herder, and the Birth of Anthropology* (Chicago: University of Chicago Press, 2002).

183. Buffon's "Variétés dans l'espèce humaine" was over 150 pages long. Although it was included at the very end of the third volume of the *Histoire Naturelle* in 1749, Kästner translated only the first two volumes for the 1750 German edition; the contents of volume 3—including Buffon's "Verschiedene Gattungen in dem menschen Geschlechte"—appeared in German in 1752.

184. While Leibniz's explicit discussion of *analysis situs* is readily available today (e.g., Loemker's translation of Leibniz's *Philosophical Papers and Letters*, 1:390; see also 1:382), it is hard to say whether anyone actually read it in the eighteenth century. Louis Dutens's well-regarded edition of Leibniz's complete works appeared only in 1768 and included neither Leibniz's essay nor his letter to Huygens from the same period, a letter that also made mention of a new geometry of situation (Leibniz to Huygens, September 8, 1679, in *Philosophical Papers and Letters*, 1:381–390). Vincenzo De Risi traces the impact of Leibniz's rumored geometry in *Geometry and Monadology*. According to De Risi, "It was thanks to Wolff that the entire scientific world first learned about *analysis situs* and attached it to Leibniz's fame," for in volume 1 of Wolff's *Elementa matheseos universae* (Halle, 1713–1715), Wolff had described Leibniz's "new analysis of situation, constructed upon a peculiar kind of calculus (which he calls calculus of situation), completely different from the calculus of magnitudes." *Geometry and Monadology*, 94–95; see also 105n104.

185. "Of the Expansion, Growth, and Delivery of the Foetus," chap. 11 of *History of Animals*, in *Buffon's Natural History*, trans. Barr, 3:263. Setting up his own discussion of "incongruent counterparts," Kant distinguished himself from Buffon in terms of the latter's identification of one dividing line: "I shall call a body which is exactly equal and similar to another, but which cannot be enclosed in the same limits as that other, its *incongruent counterpart*. Now, in order to demonstrate the possibility of such a thing, let a body be taken consisting, not of two halves which are symmetrically arranged relatively to a single intersecting plane, but rather, say, a human hand. From all the points on its surface let perpendicular lines be extended to a plane surface set up opposite to it." (2:382).

186. Jill Vance Buroker describes Kant's later discussions of incongruent counterparts in light of his subsequent account of the transcendental ideality of space in *Space and Incongruence: The Origin of Kant's Idealism* (Dordrect: D. Reidel, 1980). Her discussion of Kant's 1768 essay (pp. 50–68) concentrates—as does much of her book—on a comparison between Leibniz's and Kant's views. The majority of commentators, taking their cue from Kant's comments on Leibniz's *analysis situs* and the reference to "German philosophers," have taken the essay to be primarily in dialogue with the Leibnizian theory of space. I take some historiographical revision to be in order for such interpretations, not least because of the unlikelihood that Kant would have had any direct knowledge of Leibniz's geometry of situation. Vincenzo De Risi's brief remark regarding Buffon and Kant is a welcome exception. See *Geometry and Monadology*, 291.

CHAPTER FOUR

187. "*Subreptio*," or "*Erschleichung*" in German, was initially used in a juridical context, referring to the practice of concealing important facts in order to gain an advantage. Christian Wolff adapted the term to indicate when mere inferences were represented as actually grounded in sensation. Kant changed the meaning of "surreptitious concepts"—or "subreptive axioms," as he referred to them in the *Inaugural Dissertation* (1770)—once more, this time to indicate the illicit application of sensible concepts to metaphysical objects. The meaning of subreption had to be changed by Kant again as the contours of transcendental idealism fell into place, given that it crucially relied on a connection between sensible intuition and intellectual concepts. By 1781 "transcendental subreption" was therefore reformulated as a problem referring to the application of intellectual concepts to objects of what would have to be an intellectual intuition, a problem defining the challenges faced in the *Critique of Pure Reason*'s transcendental dialectic (e.g., A389, A509/B537, A583/B611, A619/B647, A643/B671, but note the older cast of subreption in terms of the barrier between sense and intellect at A294/B350). Intellectual concepts in the *Critique* would be limited to an "empirical employment," with the "transcendental employment" of these concepts yielding only "a logic of illusion" (A131/B170). For an extensive consideration of the role played by subreption in eighteenth-century philosophy as a whole, and in Kant in particular, see Hanno Birken-Bertsch's *Subreption und Dialektik bei Kant: Der Begriff des Fehlers der Erschleichung in der Philosophie des 18. Jahrhunderts* (Stuttgart: Frommann-Holzboog, 2006). Michelle Grier traces Kant's account of subreption from the precritical period to his discussions in the transcendental dialectic of the *Critique of Pure Reason* in *Kant's Doctrine of Transcendental Illusion* (Cambridge: Cambridge University Press, 2001).

188. Kant's newfound conviction is even more pronounced in his response to Moses Mendelssohn's comments on *Dreams of a Spirit-Seer*: "The upshot of all this is that one is led to ask whether it is intrinsically possible to determine these powers of spiritual substances by means of a priori rational judgments. This investigation resolves itself into another, namely, whether one can by means of rational inferences discover a *primitive* power, that is, the primary, fundamental relationship of cause to effect. And since I am certain that this is impossible, it follows that, if these powers are not given in experience, they can only be invented" (10:72).

189. As Kuehn describes it, "In 1766, when Green was on business in England, Scheffner wrote to Herder: 'The *Magister* [Kant] is now constantly in England, because Rousseau and Hume are there, of whom his friend Mr. Green sometimes writes to him.'" *Kant*, 155. See also Kuehn's note regarding the lack of any such letters from Green in Kant's correspondence (463n38).

190. Locke described this piece in a letter to Molyneux as a chapter meant to be added to the fourth edition of the *Essay*. Locke died before it was complete, but it was published posthumously under the title *Of the Conduct of the Human Understanding* (W. Bowyer, 1706). Kypke translated this together with a second posthumous piece from the same collection, Locke's "A Discourse of Miracles": *Johann Lockens Anleitung des menschlichen Verstandes zur Erkenntnis der Wahrheit* (Königsberg: J. H. Hartung, 1755). Alois Winter suggests that Kypke's translation might have been undertaken at Kant's urging. "Selbst Denken-Antinomien-Schranken Zum Einfluß des späten Locke auf die Philosophie Kants," *Aufklärung*, 1 (1986): 27–66. For more on Kypke's relationship to Kant, see Kuehn, *Kant*, 110–111.

191. In 1741 G. H. Thiele published a revised version of the first Latin translation from 1701 as *Johannis Lockii Armigeri Libri IV de Intellectu Humano* (Leipzig: Theophilum Georgi, 1741). Although a German translation of Locke's *Essay* was available in 1757, Kant typically paraphrased Locke in Latin when referring to him in his notes, so it is likely that he was using Thiele's translation. Further discussion of this is in Winter, "Selbst Denken-Antinomien-Schranken," 27–28; and Reinhard Brandt, "Materialien zur Entstehung der *Kritik der reinen Vernunft* (John Locke und Johann Schultz)," in *Beiträge zur Kritik der reinen Vernunft, 1781–1981*, ed. I. Heidemann and Wolfgang Ritzel (Berlin: De Gruyter, 1981), 37–68.

192. During this time period Herder, for example, reported Kant's references to Locke—references displaying an especially careful reading of Locke's account of space and personal identity—during Kant's lectures on metaphysics from 1762 to 1764 (28:39, 43). Kant's notes from 1764–1766 also considered Locke's distinction between analytic and synthetic judgments (17:278). Thirty years later Locke remained a presence in Kant's courses; as Kant characterized him in one late lecture, "Leibniz and Locke are to be reckoned among the greatest and most meritorious reformers of phi-

losophy in our times. The latter sought to analyse the human understanding and to show which powers of the soul and which of its operations belonged to this or that cognition" (9:32). Altogether, Kant would directly refer to Locke over forty times in his writings.

193. Although Leibniz's *New Essays* were reviewed by Eberhard in the *Allgemeine deutsche Bibliothek* in 1766 (III, St. II, 44ff.), the book, at least according to Giorgio Tonelli, seems not to have had an immediate scholarly impact. See Tonelli, "Leibniz on Innate Ideas and the Early Reactions to the Publication of the *Nouveaux essais* (1765)," *Journal of the History of Philosophy* 12 (1974): 437–454. John Zammito, by contrast, identifies the *New Essays* as the clear source of Kant's "great light," in *Kant, Herder, and the Birth of Anthropology*, 270. With respect to determining Kant's access to Leibniz in general, greater significance, in my view, must be placed on Louis Dutens's 1768 publication of Leibniz's complete works. The Dutens edition included, for example, Leibniz's letter to Hansch, a letter in which Leibniz directly linked himself to Plato's doctrines on the origin of ideas (in *Philosophical Papers and Letters*, 2:962–967). The Dutens edition received a laudatory review in the widely read *Nova Acta Eruditorum*, a review that included lengthy descriptions of the contents of each volume (October 1768, 433–449). The following year, a long review of the second edition of J. J. Brucker's *Historia critica philosophiae* (Leipzig, 1742–1744, 1767) appeared in the *Nova Acta Eruditorum*, with the reviewer noting the significant amount of time Brucker had devoted to Leibniz and referring readers to the new Dutens edition (April 1769, 156–173). Of Brucker's interpretation Tonelli writes, "The great historian of philosophy, J. J. Brucker (of Thomasium extraction), was able to notice what other people had overlooked, and what they would continue to overlook for a long time thereafter. He quoted extensively from Leibniz's *Epistola ad Hanschium* [the letter to Hansch], reproducing the crucial passage on innateness. Moreover, referring to the *Causa Dei*, he shows that he is aware of the core of the problem. His carefulness in doing this was probably accompanied by a certain relish in the implicit revelation of the divergency between Leibniz and Wolff." "Leibniz on Innate Ideas," 446. Note that Tonelli gives incorrect information regarding the location of Brucker's account: Leibniz's letter to Hansch is discussed by Brucker in vol. 4, p. 375. While it is uncertain whether Kant read the review of Brucker, he was certainly familiar with Brucker's discussions of Plato (and thus presumably of Leibniz, given that Brucker followed Leibniz in identifying their positions), for Kant made direct reference to Brucker when considering the connection between Plato's notion of the archetypes and the role assigned by Kant to the "Ideas" of reason in the *Critique of Pure Reason* (A316/B372). Brucker first rehearses Plato's position in his sections devoted to Greek philosophy 1:627–727; the pagination is the same in both the 1742 and 1767 editions.

194. Leibniz, preface to *New Essays on Human Understanding*, 47–48.

195. "Non tanquem conceptus *connati*, sed e legibus menti insitis (attendendo ad eius actions occasione experientiae) abstracti, adeoque *acquisiti*" (2:395).
196. I treat Kant's theory of intuition more fully in "Intuition and Nature in Kant and Goethe," *European Journal of Philosophy* 19, no. 3 (Fall 2011): 431–453.
197. Note that "cause" is now listed as an intellectual concept, whereas it was earlier listed as one of the primary cases of a sensible concept's surreptitious application to intelligible objects. This switch must be attributed to Hume, and Kant's new—if not yet fully worked out—interest in "protecting" the concept from the problem of induction. On Hume's influence with respect to this, see especially Lewis White Beck, *Essays on Kant and Hume* (New Haven, CT: Yale University Press, 1978).
198. J. de Castillon's 1770 essay was awarded the prize: "Descartes et Locke conciliés," in *Nouveaux Memoires de l'Académie Royale des Sciences et Belles-Lettres* (Berlin: Voss, 1772), 277–282. The academy had already published a similarly themed essay by N. Beguelin, "Conciliation des idées de Newton et de Leibnitz sur l'espace et le vuide [sic]" (1769), in *Histoire de l'Académie Royale des Sciences et des Belles-Lettres de Berlin* (Berlin: Haude et Spenner, 1771), and Castillon opened his piece with reference to Beguelin's work.
199. "Letter to Hansch on Platonic Philosophy," July 25, 1707, in *Philosophical Papers and Letters*, 2:963.
200. Leibniz, *New Essays on Human Understanding*, 80.
201. Kant's subsequent identification of these laws with the concepts themselves would not entail a reinterpretation of the concepts as innate; on the contrary, the laws themselves would be ultimately understood by Kant to be part of the "epigenesis of reason" itself (B167). I will discuss the stages of Kant's development of this position first in terms of the need to establish reason's unity apart from any specific laws for connecting representations (chapter 5) and second in terms of Kant's portrait of reason in the *Critique of Pure Reason* (chapter 7).
202. A reprint of Baumgarten's text is included in the academy volume devoted to the notes Kant made in his own copy of the text. See 17:5–226. All of Kant's notes made within Baumgarten's text are identified in terms of their location and arranged according to their supposed chronology, such that, for example, Kant's various remarks on §§770–775, "Origo Animae Huminae," can be traced throughout Kant's career. Since Kant taught this text every year, determining the chronological sequence of any notes made for a given section is necessarily imprecise in that it can rely only upon placement, ink color, and so on. The academy edition's two volumes devoted to Kant's notes on metaphysics (vols. 17 and 18)—including numerous pieces written on so-called loose sheets—follow Erich Adickes's dating system, a system explained by Adickes at the start of the volumes devoted to Kant's notes, marginalia, and assorted *Nachlaß* (14:lx–lxi). Adickes's system is almost always followed by the Cambridge edition of Kant's notes, though the

NOTES TO CHAPTER 4

editors often suggest longer possible time frames for a given text. Translations are here taken from the Cambridge edition wherever possible. See *Immanuel Kant: Notes and Fragments*, trans. Paul Guyer, Curtis Bowman, and Fred Rauscher (Cambridge: Cambridge University Press, 2005).

203. Kant's elaboration of the epigenesist alternative can be compared to the relatively brief remarks—at least so far as Herder recorded them—when discussing this section of *Metaphysica* in 1762–1763. See 28:889.

204. Discussing the same passage in Baumgarten thirty-three years later, for example, Kant continued to use the term "epigenesis" in contrast to the preexistence theory of origin, but in place of his concern with the physical process of blending—in fact, in place of any consideration of biological generation at all—Kant focused on the Aristotelian-derived account of "concreationism" in Baumgarten's text, rejecting this option on principle, given the soul's nature as simple substance. In language deliberately borrowed from chemical analyses, Kant here characterized the soul as either an "educt"—a thing that preexisted its new form—or as a "product," something newly produced via epigenesis. The latter theory was completely impossible, according to Kant, because a noncomposite substance like the soul could not be expected to transfer a part of itself to its offspring (28:684—these comments are taken from student lecture notes, "Metaphysics Dohna," from Kant's metaphysics course in 1792–1793). Kant made additional notes for this passage, rejecting the soul's epigenesis because of its immateriality (18:190) and its immortality (17:672, 18:429). Kant also considered the epigenesis of the soul separately in terms of a potential transfer of good or bad character (23:106–107).

205. Charles Darwin, *The Autobiography of Charles Darwin, 1809–1882, with Original Omissions Restored*, ed. N. Barlow (London: Collins, 1958), 120.

206. For more on Lambert's influence on Kant, see Lewis White Beck, "Lambert and Hume in Kant's Development," in *Essays on Kant and Hume* (New Haven, CT: Yale University Press, 1978), 108ff.; and Alison Laywine, "Kant in Reply to Lambert on the Ancestry of Metaphysical Concepts," *Kantian Review* 5 (2001): 1–48.

207. Lambert took the word *Schein* from the study of optics, and, impressed by the methods used by astronomers, he called his phenomenology a "transcendent optics" to maintain its sense as a study of appearance in all of its modes, namely, as both illusion, or *bloßes Schein*, and as real appearance or that which is indexed to the real. See J. H. Lambert, *Neues Organon* (1764), in *Gesammelte Philosophische Schriften* (Hildesheim: Olms, 1965), IV, §4. For a short discussion of this see Lewis White Beck, *Early German Philosophy: Kant and His Predecessors* (Cambridge, MA: Harvard University Press, 1969), 407–411.

208. The first two cases would be reworked for Kant's discussion of the paralogisms and the antinomies, respectively, in the *Critique of Pure Reason*.

209. There is one other instance in the *Dissertation* that similarly falls afoul of Kant's own restriction regarding any contact between intellect and sense. It occurs in Kant's discussion of "number" as an intellectual concept requiring sensible intuition for its "actualization": "Pure mathematics considers space in geometry, time in pure mechanics. To these there is added a certain concept which, though itself indeed intellectual, yet demands for its actualization in the concrete the auxiliary notions of time and space (in the successive addition and simultaneous juxtaposition of a plurality), namely, the concept of number, treated of by arithmetic" (2:397). In the *Critique of Pure Reason*, number will no longer be treated as an intellectual concept, but its connection to space and time will be essentially similar so far as it functions in the "Axioms of Intuition" as a means for "actualizing" or connecting the intellectual categories of "Quantity" to sensible intuition (A160/B199–A167/B207).
210. Herz had published his response to Kant's *Inaugural Dissertation*, sending a copy to Kant in 1771; it was in reply to this that Kant ostensibly wrote to Herz in February of 1772. See Marcus Herz, *Betrachtungen aus der spekulativen Weltweisheit* (Hamburg: F. Meiner, 1990). For more on Herz's *Betrachtung* see Eric Watkins, "The 'Critical Turn': Kant and Herz from 1770–1772," *Proceedings of the Ninth International Kant Congress* 2 (2001): 69–77.
211. Kant had already outlined the problem for himself in similar terms: "How does it come about that to that which is but a product of our self-isolated mind there correspond objects and that these objects are subject to those laws which we prescribe to them?" (17:564 and again at 17:615).
212. There are interpretive disputes regarding Kant's letter to Herz, in this instance, regarding the nature of the "objects" themselves: are the intellectual concepts meant to connect to *intellectual* objects or to *sensible* objects of experience? I take Kant to be referring only to sensible objects here, a position I discuss more fully in "The Key to all Metaphysics: Kant's Letter to Herz, 1772," *Kantian Review* 12, no. 2 (2007): 109–127.
213. Erich Adickes connects Kant's discussion here to a similar rehearsal of the various parties. In a note Adickes takes to have been written at the same time as Kant's letter to Herz—same script, similar shade of ink (Kant mixed his own ink, so different shades mark different batches and therefore different time periods)—Kant runs through the cast again, considering Crusius's "*Methodus cognitionis praestabilitae*" and the contrasting resources for ideas, "*per epigenesin vel per praeformationem*" (17:554). Here Kant pulled together the separate vocabularies of origin and connection, and he would later return to it again (a different ink color) to add yet a third layer of description in terms of dogmatic versus skeptical approaches to metaphysics. As a sign of Kant's own progress, we can read his final entry regarding this: "Alle haben die Metaphysik dogmatisch, nicht critisch tractirt" (17:554).

CHAPTER FIVE

214. Kant's presentation of the deduction would be subsequently criticized for its "obscurity," a charge to which Kant responded by repeating the distinction between the respective tasks of the objective and subjective portions of the discussion, insisting that the latter portion, while important for explaining *how* the intellectual concepts inform experience, was less necessary than the objective proof *that* they in fact did so. Kant called the achievement of a subjective deduction "meritorious," therefore, but inessential in comparison to the importance of the objective portion of the deduction (4:474). Kant later complained that even this attempt at a clarification of his goals regarding the deduction had been misunderstood, emphasizing once more the greater need to demonstrate the objective validity of concepts derived from a nonempirical origin (8:184). Such protestations aside, Kant in fact secured the rightful application of the concepts to experience—a rightfulness whose demonstration was required in order to show their "objective validity"—by way of an account of the pure understanding as the actual site of the categories' nonempirical origin and thus indeed on the *subjective* portion of the deduction. I return to this point in chapter 7.

215. In the following years Kant would repeatedly promise Herz that the "critique of metaphysics" would soon be forthcoming, complaining variously of "one major object that, like a dam, blocks" him (10:199) or of a "stone" blocking his path (10:124) when explaining its delay. The most comprehensive account of Kant's work to develop the objective and subjective strands of the argument is given by Wolfgang Carl in "Kant's First Drafts of the Deduction of the Categories," in *Kant's Transcendental Deductions: The Three Critiques and the Opus Postumum*, ed. Eckart Förster (Stanford, CA: Stanford University Press, 1989), 3–20, and in his monograph *Der schweigende Kant: Die Entwürfe zu einer Deduktion der Kategorien vor 1781* (Göttingen: Vandenhoeck and Ruprecht, 1989).

216. As Kant explains it at one point, "Since space and time are only conditions of appearance, there must be a *principium* of the unity of pure reason through which cognition is determined without regard to experience" (17:704). This is clearly broader than Kant's previous announcement regarding his "possession of a principle that will completely solve what has hitherto been a riddle and that will bring the misleading qualities of the self-alienating understanding under certain and easily applied rules" (10:144).

217. See also, "There must be two sorts of principles of unity *a priori*. Unity of the intellection of appearances *a priori*, insofar as we are determined through them, and unity of the spontaneity of the understanding, insofar as the appearances are determined through it. . . . Unity of reason. Unity of the self-determination of reason with regard to the manifold of the unity of rules or principles. Not of the exposition, i.e., of the analytic unity of ap-

pearances, but of the determination (comprehension), i.e., of the synthetic, through which the manifold as given in general (not merely to the senses) necessarily has unity" (17:707–708). In these notes Kant is inconsistent in identifying distinctive tasks for the understanding when compared to reason, but he does already think that the principles of the former's unity depend on reason (17:709).

218. Kant had previously published course announcements on four separate occasions (1757, 1758, 1759–1760, 1765–1766), and the 1775 announcement, "Of the Different Races of Human Beings to Announce the Lectures on Physical Geography of Immanuel Kant, Professor Ordinarius of Logic and Metaphysics," would be his last. The Berlin Academy edition of Kant's course announcement offers an amalgamation of two editions. The opening and closing paragraphs directly concern details of the course and are from 1775; the body of Kant's piece, however, comes from the 1777 edition. Kant had prepared the separate, expanded version of the essay for inclusion in J. J. Engel's *Der Philosoph für die Welt* (Leipzig, 1777), part 2, pp. 125–164. Although Kant was invited by the publisher and book merchant Johann Brietkopf to prepare a separate treatment on the subject of race for inclusion in an anthology, Kant declined in 1778, explaining that his "views would have to be expanded and the play of races among animals and plant species considered explicitly, which would require too much attention from me and necessitate new and extensive reading rather outside my field, since natural history is not my specialty but only a hobby and my principle aim with respect to it is to use it to extend and correct our knowledge of mankind" (10:230). Of course, Kant lectured on the "physical geography" aspect of his "hobby" some forty-eight times between 1756 and 1796, behind only logic and metaphysics in number of times taught. This demonstrated not only Kant's long-standing interest in natural history but the manner in which he considered his speculative contributions to the field to be scientifically significant and not, therefore, the mere musings of an enthusiastic hobbyist. John Zammito describes the extent to which Kant would work to establish and defend his anthropological views against Herder's position during the 1760s and 1770s in *Kant, Herder, and the Birth of Anthropology*. Klaus P. Fischer describes the Lockean basis of the position held by the philanthropinists in "John Locke in the German Enlightenment," *Journal of the History of Ideas* 36, no. 3 (1975): 438.

219. Buffon, "The Natural History of Man," sect. 9, "Of the Varieties of the Human Species," (1749), in *Natural History, General and Particular*, 3rd ed., trans. W. Smellie, (London: A. Strahan and T. Cadell, 1791), 3:57. Although Buffon published the first three volumes of his natural history together in 1749, Kästner translated and published only the first two volumes in German in 1750; volume 3 appeared in 1752. It should be noted that Buffon moved easily between "variety" (*variété*) and "race" (*race*) when discussing the matter. Since the thirteenth century, *"de bonne race"* had been used in

France when referring to the aristocracy, with a subsequent broadening of its usage in the sixteenth century in terms of a struggle between *"la noblesse d'épée"* and *"la noblesse de robe."* By the end of the seventeenth century, race was defined by the dictionary of the French Academy in a manner that made it interchangeable with the human race: *"la race mortuelle, pour dire, le genre humain."* François Bernier was thus unusual for the time, given his use of race as a means for discriminating between peoples he had encountered on his voyages. See "Nouvelle division de la terre, par les différentes espèces ou races d'hommes qui l'habitent," *Journal des sçavans* (1684): 133–140 reprinted 1685, pp. 148–155. Antje Sommer and Werner Conze's entry on "Rasse" in *Geschichtliche Grundbegriffe: Historisches Lexicon zur politisch-sozialen Sprache in Deutschland*, ed. Otto Brunner, Werner Conze, and Reinhart Koselleck (Stuttgart: Klett-Cotta, 1984), 137–141, traces some of this history. A helpful discussion of Buffon on race is in Phillip R. Sloan, "The Idea of Racial Degeneracy in Buffon's *Histoire Naturelle*," *Studies in Eighteenth-Century Culture* 3 (1973): 293–321.
220. Maupertuis, *The Earthly Venus* (1745), 76. Compare Kant's comment from a late lecture on physical geography on the general difficulties facing breeders regarding color: "Moors can give birth to white children, just as happens time to time when a white raven, crow, or blackbird is born" (9:314).
221. Ibid., 71.
222. Ibid., 78.
223. Ibid., 80.
224. See "The Natural History of Man," 3:132, 165, 173.
225. Like Maupertuis, Buffon took white to be the original color, although in Buffon's case this was on grounds tied to geographical considerations; as he summarized it, "White, then, appears to be the primitive colour of Nature, which may be varied by climate, by food, and by manners, to yellow, brown, and black, and which, in certain circumstances, returns, but so greatly altered, that it has no resemblance to the original whiteness, because it has been adulterated by the causes which have already been assigned." Ibid., 3:181. It must be kept in mind that degeneration was by no means akin to speciation in either its transformist versions espoused by Lamark or the transmutationist position held by Charles Darwin. Thus Buffon took it to be at least theoretically possible for any variety to resume its original form once it had returned to its point of geographical origin, though he thought that it would take numerous generations for the molds to be reaffected and that it would be nearly impossible to determine with certainty the specific geographical site upon which any given species had arisen.
226. Buffon, "On Degeneration" (1766), in *Natural History, General and Particular*, 3rd ed., trans. Smellie, 7:396. Kästner's translation of this entry, "Von der Abartung der Thiere," appeared in 1772.
227. Ibid., 392.

228. Ibid., 393.
229. Ibid., 394.
230. "This is very practical," Kant noted, "for in the first case [Animalculism] a man has to look closely at the character of the wife and her race [*Race*], in the second [Ovism] he does not, rather the wife has to look at the race [*Race*] of her husband. According to epigenesis one must look to both: 1) as a result of the alternative, 2) as a result of the blending [*Mischung*]. With preexistence you do not have to" (17:416; see also the continued attitude toward preexistence theory in Kant's later discussion of Mary's immaculate conception, 6:80n). Kästner's translation of Buffon's *"race"* was *"Rasse,"* which is not to suggest that Kant was reading Buffon in French, only to note the difference. Kant used *"Race"* over *"Rasse"* in both of his essays, "Von den verschiedenen Racen der Menschen" (1775), and "Bestimmung des Begriffs einer Menschenrace" (1785).
231. In his lectures on metaphysics during this period Kant would therefore insist that questions regarding animal souls amounted more to reports on our "ignorance" than the revelation of "secrets known only to philosophers" and that the real "discovery here which has cost much trouble and which only a few know [is] . . . *to cognize the limits of reason and of philosophy and to comprehend* how far reason can go here" (28:274). All that could be said about animal souls, according to Kant, was that something like a principle of activity must be logically presumed to exist in animals, since matter considered on its own is inert (28:275).
232. Kant's claim regarding the status of his own anthropological work, namely, that it is to be understood only as putting forward ideas for a speculative consideration of natural history, has been largely challenged by scholars investigating the history of the idea of race. See most recently, for example, Raphaël Lagier's *Les races humaines selon Kant* (Paris: Presses Universitaires de France, 2004). Given the number of racist remarks made throughout Kant's work in this area, it has been tempting, moreover, to see Kant's discussion of race as doing something more than providing a heuristic tool, to see it rather as a scientific scaffolding upon which Kant could justify his racist attitudes. For something of this interpretation see Mark Larrimore, "Sublime Waste: Kant on the Destiny of the Races," *Canadian Journal of Philosophy* 29 (1999): 99–125. All citations from Kant's 1775 essay will be taken from "Of the Different Races of Human Beings," in *The Works of Immanuel Kant: Anthropology, History, and Education*, trans. H. Wilson and G. Zöller (Cambridge: Cambridge University Press, 2007), 84–97.
233. Peter McLaughlin discusses Kant's vocabulary at work in the proposed taxonomy in "Kant on Heredity and Adaptation," in *Heredity Produced: At the Crossroads of Biology, Politics, and Culture, 1500–1870*, ed. S. Müller-Wille and H-J. Rheinberger (Cambridge, MA: MIT Press, 2007), 277–291. Further discussion of Kant's work to provide a scientific definition of race is in Robert Bernasconi, "Who Invented the Concept of Race? Kant's Role

in the Enlightenment Construction of Race," in *Race*, ed. Robert Bernasconi (Oxford: Blackwell, 2001), 11–36. For discussion of Kant's treatment of race in connection with his views on organisms, see Susan Meld Shell, *The Embodiment of Reason: Kant on Spirit, Generation, and Community* (Chicago: University of Chicago Press, 1996). For an account of Kant's anthropological works in connection with his critical system, see Lagier, *Les races humaines selon Kant*.

234. Kant explained the unique permanence of color against subsequent climatic variation by way of the functioning of the germs themselves for this specific trait: "Only the phyletic formation can degenerate into a race; however, once a race has taken root and has suffocated [*erstickt*] the other germs, it resists all transformation just because the character of the race has then become prevailing in the generative power" (2:442). Why racial difference alone remained invariant, for Kant, emerged only implicitly in his later discussions of the moral teleology at work in providential history—essays from the mid-1780s that were drawn in part from Kant's end discussions in his anthropology lectures. There had already been intimations of the requirement for antagonism between people in the 1770s, as Kant dismissed Maupertuis's eugenics program for a "noble sort," for example, with the claim that such a plan, while "feasible," was "well prevented by a wiser Nature because the great incentives which set into play the sleeping powers of humanity and compel it to develop all its talents and to come nearer to the perfection of their destiny, lie in the intermingling of the evil with the good" (2:431).

235. Paul Menzer, by contrast, argues that the conceptual strategy behind Kant's appeal to germs and dispositions in the anthropological writings parallels the strategy at work in Kant's early cosmological treatise (1755) regarding the formation of planets. See *Kants Lehre von der Entwicklung in Natur und Geschichte*, 106.

236. Kästner translated Buffon's "*développer*" with the same word, "*Auswickelung*," that was used in accounts of the unfolding of preformed embryos by preexistence theorists. Kant follows this terminology in 1763, equally rejecting the account of a "supernatural *Auswickelung*" proposed by preexistence theorists and the "natural *Auswickelung*" or development of the embryo out of parts that had been previously molded—parts that were "preformed" in this specific sense only—according to Buffon. (See my remarks in n. 176 for further comments regarding Buffon's reappropriation of *développer* for his own purposes and the difficulties this would pose for his translators.)

237. See also Herder's notes regarding this from Kant's lectures on metaphysics during the same period, 28:891–892.

238. Attention to Kant's language throughout the essay on race reveals Maupertuis, and particularly Buffon, to have been Kant's central interlocutors on race at this juncture. Their specific accounts of generation seem, more-

over, to have been Kant's primary models when considering the role of germs and dispositions in his own theory. Thus while it can be supposed that Kant was aware of Bonnet's account of preexistent germs as the most nuanced defense of preexistence theory, there are only a very few direct references to Bonnet made by Kant: once in 1768 in reference to Bonnet's thought experiment (originally Condillac's) regarding the senses—the so-called *bonnetische Statue*—and three times in the 1780s while criticizing Bonnet's notion of the "great chain of being" (2:381, B696, 8:180n.). The key words to trace here in Kant's usage are the nouns and derivatives of *auswickeln* (to unwrap, usually translated as "unfold" in English), and *entwickeln* (to develop). On the whole, Buffon's German translators used *auswickeln* for *développer* during Buffon's discussions of generation, whereas Bonnet's translators tended to use *entwickeln* for *développer*, particularly in his *Palingénésie* (1770; vol. 1 trans. 1770, vol. 2 trans. 1772). While there is still some mixing of the terms in the German translation of Bonnet's *Contemplation de la nature* (1764–1766; trans. 1766)—such that there is reference, for example, to the "universal laws of unfolding [*Auswickelung*]" by which "a new head develops [*entwickelt*]" on a hydra—in Bonnet's book on palingenesis there are almost no occurrences of the terms *Auswickelung* (0) or *auswickeln* (4), compared to *Entwickelung* (21) or *entwickelt* (25). (By comparison, in *Betrachtung der Natur*: *Auswickelung* occurs 9 times, *auswickeln* 4, *Entwickelung* 15, and *entwickelt* 19, but note that *auswickeln* is also used as a translation of *Evolution*, not only *développer*, in this text.) For references to Bonnet see *Betrachtung über die Natur* (Leipzig: J. F. Junius, 1766); and *Philosophische Palingenesie*, trans. J. C. Lavater, 2 vols. (Zurich: Orell, Gessner, Fuesili, 1770). These differences suggest that Kant's own pattern of word choice reveals, to some extent, his linguistic influences. Up until 1777 Kant used *Auswickelung* some 14 times, compared to 11 uses *Entwickelung*. After 1777 Kant used *Auswickelung* 9 times, but *Entwickelung* would appear some 112 times. What can perhaps be seen here is that the change in Kant's usage from *auswickeln* to *entwickeln* after 1777 demonstrates that he was reading Buffon, not Bonnet, when appealing to germs and dispositions in his discussion of race in 1775. After 1777, after, that is, Kant had carefully read through volume 2 of Tetens's *Versuche*, where there were literally hundreds of pages devoted to a discussion of generation in terms of the three positions Tetens sought to combine—Buffon's "concreationist" account, Caspar Wolff's theory of epigenesis, and Bonnet's system of the *Entwickelung* of preexistent germs—Kant would switch almost exclusively to the use of *Entwickelung*. Tetens's own position argued for an "Evolution through Epigenesis." See J. N. Tetens, *Philosophische Versuche über die menschliche Natur und ihre Entwickelung* (Leipzig: Weidmanns Erben und Reich, 1777), 2:508–513. Amid all this data, however, there are two points that should not be forgotten: first, that Buffon explicitly rejected preexistence theories despite his use of *développer* and, second, that although Bonnet rejected the notion of preformed, miniature individuals, he was indeed an advocate of preexistent structures contained in the germs of an animal, structures whose enlargement he took to be captured by his notion of *développer*. Beyond the context of Buffon and Bonnet in German translation, by the mid-1780s *Entwickelung* would come to convey an unspecified sense of activity in development, whereas *Evolution* would become linguistically established as the preexistence alternative.

239. Kant used the language of germs and dispositions throughout his writings, particularly in the moral works. Indeed, taken altogether, according to Kant there was a germ for metaphysics, for reason, for evil, for cultivation, and for enlightenment. And while Kant tended to distinguish *Naturanlagen* from *moralische Anlagen* (with the latter frequently blurring the lines between itself and *Gessinung*), undenominated *Anlagen* appeared as frequently throughout Kant's works as did germs.

240. In later lectures Kant would weaken the claim that all individuals were themselves perfectible given the universal perfectibility of the species. In the lectures from 1775–1776, for example, the "savage Indian or Greenlander" was said to have "the same germs as a civilized human being, only they are not yet developed" (25:651). But by 1781–1782 Kant suggested that in cases where no advancement had occurred in a people over time, "one must assume that there is a certain natural disposition [*Naturanlage*] within them that cannot be overcome." *Immanuel Kants Menschenkunde: Oder Philosophische Anthropologie nach handschriftlichen Vorlesungen*, ed. F. C. Starke (Leipzig, 1831), 352. And by 1790–1791, Kant was ready to say that although the point of the species' natural dispositions was to lead it to the formation of a civil society (and ultimately, thereby, a moral kingdom of ends), neither the Negro nor the American Indian would ever be capable of creating (*stiften*) such a society themselves. See *Kants Anweisung zur Menschen- und Welterkenntnis*, ed. F. C. Starke (Leipzig, 1831), 119, 121. This shift in Kant's thinking paralleled the increasing emphasis in his lectures on the role of a "moral disposition"—in this case, purported laziness, weakness of character, etc.—as an occasioning cause alongside climate and nutrition in the formation of races and varieties. Pauline Kleingeld traces a change in Kant's views after 1792 regarding the hierarchy of the races with respect to their moral characterizations, arguing that Kant was forced to weaken this aspect of his theory—while still maintaining the viability of race as a physiological concept—once he took up the question of cosmopolitan right. See "Kant's Second Thoughts on Race," *Philosophical Quarterly* 57 (2007): 573–592. Drawing primarily on Kant's lectures on physical geography, Robert Bernasconi revisits Kleingeld's claim in order to describe, by contrast, the extent to which Kant's earlier attitudes remained fundamentally intact during the physical geography lectures Kant delivered during the 1790s. See Bernasconi, "Kant's Third Thoughts on Race," in *Reading Kant's Geography*, ed. Stuart Elden and Eduardo Mendietta (Al-

bany: SUNY Press, 2011), 291–318. For Kant's public criticisms of colonialism and, in particular, the treatment of the Sugar Islands' slave population and the moral duty to educate the children of slaves, see his published remarks from the 1790s, namely, *Perpetual Peace* (8:359) and the much later *Metaphysics of Morals* (6:266, 283, 314, 331).

241. Kant used Basedow's *Methodenbuch* as a textbook when lecturing on pedagogy during the winter semester of 1776–1777. A good sense of Kant's commitment to the school during this period emerges from his letter exchanges regarding it. See esp. 10:191–195.

242. Kant's encomium was even more pronounced at the end of that semester's anthropology lectures: "The present Basedowian institutes are the first that have come about according to the perfect plan of education. This is the greatest phenomenon which has appeared in this century for the improvement of the perfection of humanity, through it all schools in the world will receive another form, and the human race will thereby be freed from the constraints of the prevailing schools" (25:722–723). The next time Kant would be willing to express such optimism regarding humanity's progress would be with respect to the feelings of sympathy generated by the French Revolution, feelings he took to be signs of a deep moral disposition (*moralische Anlage*) in the human race (7:85).

243. This claim guides every one of the nine propositions outlined by Kant in his 1784 essay "Idea for a Universal History with a Cosmopolitan Aim." See esp. 8:18–23, 25, 30.

244. Kant would make a similar argument regarding this kind of susceptibility when discussing moral feeling in the *Metaphysics of Morals*. In this discussion Kant described a "vital moral force" capable of exciting moral feeling as a nonpathological response to the mind's representation of the moral law. In his words, "No human being is entirely without moral feeling, for were he entirely lacking in receptivity to it he would be morally dead; and if (to speak in medical terms) the moral vital force could no longer excite this feeling, then humanity would dissolve (by chemical laws, as it were) into mere animality and be mixed irretrievably with the mass of other natural beings" (6:400). This susceptibility or "receptivity" was original to us, and its function was to orient the mind toward the moral law without thereby compromising its freedom. The ability to be quickened by the moral force in this manner was thus indeed innate, according to Kant, but not in the sense of its having been implanted. Moral feeling was in this sense simply "inscrutable," and rather than speculate further regarding its origins, our responsibility lay instead toward its "cultivation" (*cultiviren*) (6:400; cf. 8:344).

245. *On a Discovery Whereby Any New Critique of Pure Reason Is to Be Made Superfluous by an Older One* (1790), trans. Henry Allison, in *Theoretical Philosophy after 1781* (Cambridge: Cambridge University Press, 2002). Kant would undertake a similar approach when trying to negotiate the question of de-

NOTES TO CHAPTER 5

termining the innate character of goodness and evil in mankind (e.g., evil is innate only insofar as freedom as its a priori ground is innate) (6:32, 35).

246. J. A. Eberhard was the Leibnizian philosopher responsible for one of the more scathing attacks Kant's *Critique of Pure Reason* would receive. A thorough history of this accompanies Henry Allison's translation of Eberhard's piece. See H. Allison, *The Kant-Eberhard Controversy* (Baltimore: Johns Hopkins University Press, 1973), 1–104.

247. Leibniz, *New Essays on the Human Understanding*, 80.

248. I therefore disagree with Phillip Sloan's widely cited account of Kant's position on this point during both this period and in the first *Critique*. While Sloan is right to identify the continuity between Kant's vocabulary of germs and dispositions when discussing intellectual concepts in the 1770s and 1781 (e.g., A66/B91) the continuity stands with regard to their epigenesis, not their preformation in the Leibnizian sense of ideas implanted in the mind. See Sloan, "Preforming the Categories: Eighteenth-Century Generation Theory and the Biological Roots of Kant's A Priori," *Journal of the History of Philosophy* 40 (2002): 229–253. Claude Piché, by contrast, interprets Kant in a manner close to my own on this point, though I depart from him regarding the shift he sees in Kant toward the epigenesis of experience during the 1780s. See "The Precritical Use of the Metaphor of Epigenesis," in *New Essays on the Precritical Kant*, ed. Tom Rockmore (New York: Humanity Books, 2001), 182–200.

CHAPTER SIX

249. The clearest version of this argument—including within it a short rehearsal of Tetens's theory of cognition—is provided by Herman de Vleeschauwer in his influential account of Kant's development, *La déduction transcendentale dans l'oeuvre de Kant* (Paris: É. Champion, 1934), 1:299–315. A shorter summary of de Vleeschauwer's argument in English is in *The Development of Kantian Thought: The History of a Doctrine*, trans. A. R. C. Duncan (London: Thomas Nelson and Sons, 1962), 82–88. More recent appraisals of the differences between Tetens's and Kant's ultimate theories of representation can be found in separate discussions by Beck, Carl, and Kitcher. Beck, *Early German Philosophy*, 412–424; Carl, *Der schweigende Kant*, 119–126; and Patricia Kitcher, *Kant's Thinker* (Oxford: Oxford University Press, 2011), 31–39. The best discussion of Tetens's own theory of cognition, of its connection to Kant, and of the assessment made by both Kant's contemporaries and his successors regarding Tetens's influence on the *Critique* is in Wilhelm Uebele's *Johann Nicolaus Tetens nach seiner Gesamtentwicklung betrachtet, mit besonderer Berücksichtigung des Verhältnisses zu Kant* (Berlin: Reuther and Reichard, 1911), esp. chap. 3, 69–156. A brief selection of passages from the first volume of Tetens's *Philosophische Versuche über die menschliche Natur und ihre Entwickelung* is available in English translation from Eric Watkins

in his *Kant's Critique of Pure Reason: Background Source Materials* (Cambridge: Cambridge University Press, 2009), 353–391.
250. This aspect of Lambert's thought is also described by Beck in *Early German Philosophy*, 402–412.
251. J. N. Tetens, *Einleitung zur Berechnung der Leibrenten und Anwartschaften*, 2 vols. (Leipzig: Weidmanns Erben und Reich, 1785-1786).
252. Kant would complain about Tetens's lack of attention to the first *Critique* on three separate occasions, describing Tetens as one of "the only men through whose cooperation this subject could have been brought to a successful conclusion before too long, even though centuries before this one have not seen it done. But these men are leery of cultivating a wasteland that, with all the care that has been lavished on it, has always remained unrewarding. Meanwhile people's efforts continue in a constant circle, returning always to the point where they started; but it is possible that materials that now lie in the dust may yet be worked up into a splendid construction" (10:341). Kant also mentions Tetens in this vein at 10:346. Given these remarks, one might wonder at the lack of published attention on Kant's part to Tetens's *Versuche*—at least prior to Kant's being attacked for his "psychologism"—here Uebele's claim regarding Kant's subsumption of Tetens under Locke as the universal representative of empiricist approaches seems right. See Uebele, *Johann Nicolaus Tetens nach seiner Gesamtentwicklung betrachtet*, 188.
253. See J. N. Tetens, *Über die allgemeine spekulativische Philosophie* (Bützow: Berger und Boednerschen Buchhandlung, 1775). Uebele describes the specific nature of the *Dissertation*'s influence on the 1775 piece—an influence operating primarily through Kant's theory of spatial intuition. Ibid., 103ff.
254. Although Tetens intimates his ultimate position in the preface to volume 1 of the *Philosophische Versuche*, the specific account of the manner in which he sees an analogous balance between form and force to be operating in the generation of the body, the brain, and the soul, is first introduced in the second volume, 2:508–513.
255. Ibid., 1:vii–ix, xiv. It is worth remembering here that psychology first emerged in the eighteenth century as a method meant to correct the excesses of speculative metaphysics. This background makes sense of Bonnet's insistence, for example, that he not be compared to Leibniz, since he took his psychological investigations into cognition to have nothing to do with those of the older metaphysician. Thus in the preface to the second German edition of Bonnet's *Palingenesie* the physiognomist Johann Caspar Lavater ruefully included long excerpts from Bonnet's letter to Lavater admonishing him for having included misleading remarks concerning Leibniz and Bonnet in his German translation of the first edition. See Charles Bonnet, *Philosophische Palingenesie oder Gedanken über den vergangenen und künftigen Zustanden lebender Wesen*, 2nd. ed., trans. J. C. Lavater (Zurich:

NOTES TO CHAPTER 6

Orell, Gessner, Fuesili, 1770), 1:xi, xiii. Tetens was similarly concerned to position his work in 1777 as distinct not only from traditional metaphysics but also from "the analytic or anthropological method" being used by the "new psychology." See Tetens, *Philosophische Versuche*, 1:iv–v. Martin Kusch provides a thorough history of the rise of attacks on psychologism in the wake of Kant and Hegel in nineteenth-century Germany in *Psychologism: A Case Study in the Sociology of Philosophical Knowledge* (London: Routledge, 1995). A recent anthology devoted to this history is also helpful, especially the introductory essay: Dale Jacquette, ed., *Philosophy, Psychology, and Psychologism: Critical and Historical Readings on the Psychological Turn in Philosophy* (Dordrecht: Kluwer Academic, 2003).

256. Tetens, *Philosophische Versuche*, 1:v.
257. Ibid., xi.
258. Hamann to Herder, May 17, 1779, cited by Arnulf Zweig in *Correspondence*, The Cambridge Edition of the Works of Immanuel Kant (Cambridge: Cambridge University Press, 1999), 168n1.
259. A. G. Baumgarten, *Metaphysica*, 3rd ed. (Halle: C. H. Hemmerde, 1757). A reprint of Baumgarten's text is included in the academy volume devoted to the notes Kant made in his own copy of the text. See 17:5–226. Thomas Sturm discusses Baumgarten's account of empirical psychology in relation to Kant in "Kant on Empirical Psychology," in *Kant and the Sciences*, ed. Eric Watkins (Oxford: Oxford University Press, 2001), 163–184.
260. Rudolf Makkreel considers the impact of Kant's views on empirical psychology on the *Critique of Judgment* in "Kant on the Scientific Status of Psychology, Anthropology, and History," in *Kant and the Sciences*, ed. Eric Watkins (Oxford: Oxford University Press, 2001), 185–201.
261. This kind of distinction lay behind Kant's later taxonomy regarding the metaphysics of morals as well, insofar as moral empiricism was similarly determined to be a species of "practical anthropology" for its consideration of the will only so far as it could be sensuously affected (4:388), and it grounded Kant's insistence that "we should not dream for a moment of trying to derive the reality of the basic moral principle from the special characteristics of human nature" (4:425).
262. I follow Wolfgang Carl in dating the "Metaphysics L1" lectures to the winter semesters between 1777 and 1778 and between 1779 and 1780. See *Der schweigende Kant*, 117–118. The editors of the Cambridge edition of Kant's lectures provide the history of attempts to determine the precise dating of these lectures in *Lectures on Metaphysics*, trans. and ed. Karl Ameriks and Steve Naragon (Cambridge: Cambridge University Press, 1997), xxxi–xxxiii.
263. In Kant's anthropology notes from this period he also identifies the genius as having a faculty of *"Urbildung,"* which Rudolf Makkreel describes as the capacity to form archetypes: "Das Urbildende Talent ist genie, das Nachbildende nicht" (15:232). Makkreel discusses Kant's early account of image

formation and the imagination in connection to his later discussions in the *Critique of Pure Reason* and the *Critique of Judgment* in *Imagination and Interpretation in Kant: The Hermeneutical Import of the "Critique of Judgment"* (Chicago: University of Chicago Press, 1990).

264. This use of the imagination in the formation of "determinate" judgments is connected to logical subordination in a manner that is distinct from the imagination's role in "reflective" judgments, where aesthetic coordination (versus logical subordination) marks the formative process. This difference is already anticipated at 15:131 and in Kant's notes in his copy of Meier's logic at 16:119. Kant used Georg Friedrich Meier's logic textbook, *Excerpts from the Doctrine of Reason*, for his logic courses. See Meier's *Vernuftlehere* (Halle: J. J. Gebauer, 1752) and *Auszug aus der Vernunftlehre* (Halle: J. J. Gebauer, 1760). It is worth recalling Beck's comment on Meier, since, according to Beck, Meier went beyond Baumgarten "in recognizing the role of imagination in all intellectual activity, even in the formation and application of concepts in the process of knowing." *Early German Philosophy*, 286.

265. See Tetens, *Philosophische Versuche*, 1:104–127, 159–161. Thus, for example, Tetens takes a thorough "deduction" of the various aspects of the formative power as an investigation starting from experience (though he ultimately judges this to lie outside the particular purview of his first investigation): "Eine ausführliche physische Untersuchungen der bildenden Kraft der Seele, in der jede Regel, jedes Gesetz ihrer Wirksamkeit so vollkommen mit Beobachtungen belegt würde, als eine überweisende Deduktion aus Erfahrungen es erfordert, würde über die Gränzen hinausgehen, die ich mir in dem gegenwärtigen Versuch gesetzet habe" (119).

266. Although Kant would go on to describe the foundational role played by the transcendental synthesis of the imagination for any empirical synthesis according to the laws of association (e.g., A106–109) in the 1781 edition of the *Critique of Pure Reason*, he would remove all references to empirical synthesis from his account in the second (1787) edition, since by then he had decided that discussion of the "reproductive" imagination belonged in fact to empirical psychology as opposed to transcendental philosophy (B152).

267. The constraint applied to vocabulary regarding the soul is everywhere the same when referring to transcendental apperception also as "the bare I think" (e.g., A117, B132).

268. Dieter Henrich argues similarly in "Kant's Notion of a Deduction and the Methodological Background of the first *Critique*" in *Kant's Transcendental Deductions*, ed. Eckart Förster (Stanford, CA: Stanford University Press, 1989): "The unity of apperception is the origin of the system of the categories and the point of departure for the deduction of the legitimacy of their usage" (45). "In its fundamental structure," as Henrich rightly puts it, "the transcendental deduction is patterned on a deduction that aims to justify an acquired right by appealing to particular features of the origin

NOTES TO CHAPTER 7

of the categories and their usage" (39). I believe that Henrich is wrong, however, in suggesting that Kant's method of proof in the deduction is not ostensive (41). While I agree with Henrich's reading of the implicit nature of reflection and its connection to the *indirect* sense of apperception as a bare "I think," Kant explicitly identifies the transcendental deduction as a (nonsyllogistic) ostensive proof. The ostensive nature of Kant's proof in fact supports Henrich's (and my own) reading of the deduction, since it "combines with the conviction of its truth insight into the sources of its truth," or as Kant also puts it, ostensive proofs proceed by "reviewing the whole series of grounds that can lead us to the truth of a proposition by means of its complete possibility" (A789/B817, A791/B819). Note Henrich's comment regarding his widely cited earlier treatment of the transcendental deduction: "When I wrote the paper, I had no idea what a deduction consists in and took for granted that it was exhaustively defined as a chain of syllogisms. But it isn't, and after finding out that this is so, I must relativise what I said in that paper. The deduction of the second edition is indeed a proof within two steps; but Kant's main reason for separating the two steps is their distinctive contribution to an understanding of the origins of knowledge" (252). Henrich's first essay remains highly influential. See "The Proof-Structure of Kant's Transcendental Deductions," *Review of Metaphysics* 22 (1969): 640–659.

CHAPTER SEVEN

269. This comment was made by Johann Schultz, court chaplain and longtime supporter of Kant's work. Schultz's remarks were embedded in his anonymous review of a book on logic by Johann Ulrich. The specific charge of obscurity became well known in part because it was one of the very few criticisms of the *Critique of Pure Reason* to which Kant directly responded. Kant replied to Schultz's review in a lengthy footnoted remark in the preface to the *Metaphysical Foundations of Natural Science* (4:474–476), promising to make up for the difficulties in his exposition "at the earliest opportunity." Kant was given that opportunity less than a year later, as he began revisions for the second edition of the *Critique* in the summer of 1786. As for Kant's own views regarding the deduction's relative opacity, it is noteworthy that he took that task to be relatively straightforward in comparison to the far more challenging attempt at a similar deduction in the moral sphere (e.g., 5:46). Schultz's review is reprinted in Brigitte Sassen's collection of translations as "*Institutiones Logicae et Metaphysica* by Jo. Aug. Hen. Ulrich," in *Kant's Early Critics: The Empiricist Critique of the Theoretical Philosophy* (Cambridge: Cambridge University Press, 2000), 210-214.
270. In what seems to have been an outline written for this discussion in 1780, Kant had drawn up a preliminary list of the various paths that had been so far taken by reason in its metaphysical investigations. As he saw it, there

had been four altogether: "The empirical path and universality through induction. The fanatical path of intuition through the understanding. That of predetermination through innate concepts," and finally, "the *qualitas occulta* of the healthy understanding [common sense] which gives no account" (18:272). With reason's history thus laid out, Kant stopped to criticize the empiricists in particular for having followed a route that had done away with necessity in both mathematics and experience. "Thus Locke," Kant concluded, "who earned almost too much honor after Leibniz had already refuted him, falls by the wayside. There thus remain epigenesis, mystical intuition, and involution. Finally there is also the *qualitas occulta* of common reason" (18:272–273). From here Kant went on to create a second list of positions that had advanced logical systems of cognition. These systems could be either empirical or transcendental, "the former Aristotle and Locke," Kant wrote, "the latter either the system of epigenesis or that of involution, acquired or inborn" (18:275). It was clear where Kant placed himself on these lists, for as the first *Critique* would go on to show, only the critical path had discovered a system where transcendental logic emerged epigenetically from out of reason itself.

271. In the first *Critique* Kant characterized the difference between the two sections as lying between a focus on the "materials" versus the "plan" for a building, defining the "Doctrine of Method" as the section concerned with the "formal conditions of a complete system of pure reason" (A708/B736). He would repeat this distinction in the *Metaphysics of Morals*, explaining there that insofar as method concerns the "form of the science," it stands with respect to the "Doctrine of Elements" as the "ground plan of the whole" (6:413).

272. See n. 81 for a fuller explanation of growth by "intussusception."

273. Kant made the same point in the *Metaphysics of Morals*: "Since, considered objectively, there can be only one human reason, there cannot be many philosophies; in other words, there can be only one true system of philosophy from principles, in however many different and even conflicting ways one has philosophized about one and the same proposition"; only by paying attention to that fact, according to Kant, would it be possible to demonstrate the "unity of the true principle which unifies the whole of philosophy into one system" (6:207). In *Religion Within the Bounds of Reason Alone* Kant described the historical self-development of religion in a manner that was also indebted to his description of reason. For example, "we must have a principle of unity if we are to count as modifications of one and the same church the succession of different forms of faith which replace one another . . . for this purpose, therefore, we can deal only with the history of the church which from the beginning bore with it the germ and the principles of the objective unity of the true and *universal* religious faith to which it is gradually being brought nearer" (6:125). This point would be mirrored in the social and political sphere once Kant took up the

history of civil constitutions in his essay *Perpetual Peace*, a history whose epochal determinations were unified throughout, as Kant saw it, by the unfolding of reason's concept of right (8:350)—a point that Kant repeated in terms of the "evolution of a constitution" in both the *Conflict of the Faculties* (7:87, see also 7:91) and the *Metaphysics of Morals* (6:340). In his *Philosophy of Art* Schelling would mirror Kant's account of philosophy's organic development across history: "There is only *one* philosophy and *one* science of philosophy. What one calls different philosophical sciences are mere presentations of the *one*, undivided whole of philosophy under different ideal determinations. . . . The relationship between the individual parts in the closed and organic whole of philosophy resembles that between the various figures in a perfectly constructed poetic work, where every figure, by being a part of the whole, as a perfect reflex of that whole is actually absolute and independent in its own turn." *The Philosophy of Art*, trans. D. Stott (Minneapolis: University of Minnesota Press, 1989), 281–282.

274. This was not quite yet Herder's definition of *Bildung* as "rising up to humanity through culture," but Kant's notion of reason's perfectibility—under the influence of Rousseau as much as anyone else—was foundational for views in line with Herder's own. Discussions of *Bildung* as both cultural advancement and the progressive development of the species were increasingly prominent in the last decades of the eighteenth century. Thus whereas J. G. Walch's important *Philosophisches Lexikon* from the 1730s made no mention at all of *Bildung*, by the time W. T. Krug put together the next big philosophical dictionary at the beginning of the nineteenth century there was a lengthy entry. In Krug's entry *Bildung* referred to the cultivation of "head, heart, and taste" corresponding respectively to intellectual culture, morality, and aesthetics. (Although Krug had referred to the critical philosophy in many of his entries, he did not make the obvious connection to Kant's three *Critiques* regarding this three-pronged cultivation of humanity.) Krug listed a *Bildungstrieb*—without naming Blumenbach—as the correlative branch of *Bildung* with respect to natural formation and development. See W. T. Krug, "Bildung," in *Allgemeines Handwörterbuch der philosophischen Wissenschaften nebst ihrer Literatur und Geschichte*, 2nd ed., (Leipzig: F. A. Brockhaus, 1832), 1:358–360. Insofar as the concept of *Bildung* was taken to be operating at the levels of both the culture and nature, Denise Gigante sees it as the bridge between aesthetics and science, as in fact the theoretical basis of Romantic science itself. See *Life: Organic Form and Romanticism* (New Haven, CT: Yale University Press, 2009), 46. For Herder's definition see Gadamer's discussion, *Truth and Method*, trans. Joel Weinsheimer and Donald Marshall (New York: Continuum International, 2004), 10. Günter Zöller discusses the impact Rousseau's notion of perfectibility would have on Kant in "Between Rousseau and Freud: Kant on Cultural Uneasiness," in *New Studies on Kant*, ed. Pablo Muchnik (Cambridge: Cambridge Scholars, 2013).

275. This sense of metaphysics' self-development as something that has been, as Kant put it here, "wisely organized for great ends," leads Claude Piché to argue that reason is subject here to the same teleological work of "providence" that one finds described in Kant's early history essays with respect to humanity. Piché thus concludes that "if such a parallel is legitimate, this means that critical philosophy relies on ultimate metaphysical premises that are not in themselves subjected to philosophical investigation, simply because they sustain Kant's critical project from the beginning." "The Precritical Use of the Metaphor of Epigenesis,", 195. Piché's conclusion might very well be correct in general regarding Kant's reliance on metaphysical premises, but it should at least be seen that because the history of reason is in fact conceived by Kant as something that has developed out of reason itself, his appeals to "providence" in the history essays might as well be appeals to the work of reason. This conclusion is fully in keeping with Kant's own language insofar as one must then simply recognize that reason is everywhere oriented only by goals that it has set for itself; Kant is explicit on this point with respect to practical reason's appeal to divine providence in his essay *Perpetual Peace*, 8:362.
276. E.g., A309/B366, A328/B383, A336/B393, A339/B397, A778/B806.
277. As Kant put it in the *Critique of Judgment*, "Even in one and the same tree we may regard each branch or leaf as merely set into or grafted onto it, and hence as an independent tree that only attaches itself to another one and nourishes itself parasitically" (5:371–372). The clear successor to Kant in this appeal was Goethe. Demonstrating the manner in which "the whole was reflected in each of the parts" was key to all scientific investigations, according to Goethe, but plants were especially good examples of this, since "in organic formations, several identical forms can and must develop, in, with, beside, and after one another. They indicate multiplicity in unity. Every leaf, every bud, is entitled to become a tree. . . . We cannot repeat often enough that each organization unites various active parts." See "Later Studies and Collections," in *Goethe's Botanical Writings*, trans. Bertha Mueller (Woodbridge, CT: Ox Bow, 1989), 100; and "Nacharbeiten und Sammlungen," in *Goethes Schriften zur Morphologie*, ed. Dorothea Kuhn (Frankfurt: Deutscher Klassiker, 1988), 464. Since Goethe agreed with Schlegel in believing that science must become art, he would make the point in both poetry—"Asleep within the seed the power lies/Foreshadowed pattern, folded in the shell/Root, leaf, and germ, pale and half-formed"—and in prose: "Leaf and eye [root point] are inseparable: every leaf has an eye behind it, every eye has leaves which overlap like scales, and each of these leaves (the first as well as those that follow) gives us a picture of the whole plant. As a result we must think of any point on the plant as an eye with the potential to produce a root." Goethe's poem "The Metamorphosis of Plants," trans. Heinz Norden, in *The Metamorphosis of Plants*, introduction and photography by Gordon L. Miller, trans. Douglas Miller (Cambridge,

NOTES TO CHAPTER 7

MA: MIT Press, 2009), 1; the prose passage is from "Leaf and Root," in *Goethe: Scientific Studies*, trans. Douglas Miller (New York: Suhrkamp, 1983), 99; "Blatt und Wurzel," in *Goethes Schriften zur Morphologie*, 660.

278. Extended reflections on the "needs" of reason appear in Kant's 1786 piece meant to respond to Jacobi in light of the developing pantheism controversy. See "What Does It Mean to Orient Oneself in Thinking," in *Religion and Rational Theology*, trans. Allen Wood (Cambridge: Cambridge University Press, 1996), 7–18; 8:133–146. But it must be said that discussion of needs, in this sense, is present everywhere once one turns to the practical writings, particularly in the *Critique of Practical Reason*, with respect to both the moral law and reason's postulates in connections with it (e.g., 5:114, 5:125).

279. It was with this in mind that Kant criticized Hume's "geographical" approach to reason, an approach demanding a linear path when mapping its extent (A760/B788; cf. 8:135). "Our reason," as Kant put his response to Hume on this point, "is not like a plane indefinitely extended, the limits of which we know in a general way only; but must rather be compared to a sphere, the radius of which can be determined from the curvature of the arc of its surface . . . outside this sphere (the field of experience) there is nothing that can be an object for reason" (A762/B790). It bears noting, nonetheless, that Kant would go on to differentiate between the speculative attempts of theoretical reason and the subjective need, on the part of practical reason, to assert the objective reality of the intelligible objects it had postulated—a difference Kant describes as the "enigma of the critical philosophy" (5:5). Kant was clear, moreover, on the primacy of practical reason's goals in comparison to those of speculative reason, a hierarchy whose arrangement was necessary in order to avoid a conflict of reason within itself (5:121).

280. Kant's phrasing here is unusual, and the grammar in the original is ambiguous with respect to the *"Selbstgebärung"* of reason; that is, it is unclear whether this is simply an appositional clause meant to reaffirm the generation of concepts from out of reason or if Kant is indicating that reason is itself self-born (given Kant's other remarks regarding the "self-development of reason" or the "epigenesis of reason," I take the latter to be the case). Kant's English-language translators have struggled with the word *Selbstgebärung*, appealing in all cases to vocabulary taken from the life sciences. Kemp Smith uses, for example, "spontaneous generation," and Guyer-Wood chooses "parthenogenesis"; both of these translations take the clause regarding the *"Selbstgebärung"* of reason to be appositive. This is the only place Kant uses the term *Selbstgebärung*, but although it was not a word used by him in place of epigenesis elsewhere, from all of his other comments regarding epigenesis it seems clear that the "epigenesis of reason" (B167) is understood by him to mean that reason is indeed self-born. The only regular use of words similar to *Selbstgebärung* in the German of

Kant's time appear in reference to Hesiod's account of Gaia (Earth), who gave birth to her first children—sky, hills, and sea—without a father (*die ihre Kinder ohne einen Vater aus sich selbst gebären kann*). Gaia's ability to produce children "without sweet union of love" is almost always described as a case of "parthenogenesis" in English-language discussions of Hesiod on this point.

281. For Hume's use of "natural affinity," see *A Treatise of Human Nature* (1739–1740), ed. Selby-Bigge, rev. ed. ed. P. H. Nidditch (Oxford: Oxford University Press, 1978): "When the mind is determined to join certain objects, but undetermined in its choice of the particular objects, it naturally turns its eye to such as are related together. They are already united in the mind: they present themselves at the same time to the conception; and instead of requiring any new reason for their conjunction, it would require a very powerful reason to overlook their *natural affinity*" (504n71, italics added). Of course, Kant did not want the "special affinity" proposed by Leibniz either, that is, the connection between mind and idea guaranteed by the fact that each stemmed from an identically divine origin: "What makes the exercise of the faculty easy and natural so far as these truths are concerned is a *special affinity* which the human mind has with them; and that is what makes us call them innate (italics added). So it is not a bare faculty, consisting in a mere possibility of understanding those truths: it is rather a disposition, an aptitude, a preformation, which determines our soul and brings it about that they are derivable from it." Leibniz, *New Essays on Human Understanding*, 80. Henry Allison takes up the issue of affinity with a different agenda in "Transcendental Affinity—Kant's Answer to Hume," in *Kant's Theory of Knowledge*, ed. Lewis White Beck (Dordrecht: Reidel, 1974), 119–127.

282. Kant would subsequently point to reason as the birthplace of the moral law as well. Thus in the *Groundwork*, for example, Kant would explain that "it is here that she has to show her purity as the authoress of her own laws—not as the mouthpiece of laws whispered to her by some implanted sense," and also not as having received them from experience, which "would foist into the place of morality some misbegotten mongrel patched up from the limbs of a varied ancestry and looking like anything you please, only not like virtue" (4:425–426). Morality would instead have to be born from out of pure reason itself, for only that kind of pedigree could ensure its sovereignty over the will on the basis of birthright alone. This account of reason's role in giving birth to individual morality ran parallel to its work to achieve the moral advancement of the species as a whole. Perfect moral advancement would culminate in the creation of a "kingdom of ends," according to Kant, and bring with it the completion of the history of reason. This was an idea of moral perfection born out of reason itself, an idea that lay invisibly within humanity as something whose conception was "self-developing" (*sich entwickelnden*) and whose existence needed to be under-

stood as a "self-fertilizing germ" (*besamenden Keim*) of goodness in the species as a whole (6:122). It was just this aspect of Kant's philosophy that would earn harsh criticisms from Hegel, however, since he took Kant's notion of pure reason to be impotent, something capable of supplying only an empty notion of unity, that is, one that had never been lifted out of intellect by the intellectual intuition of itself. On the basis of such sterility, as Hegel saw it, Kant could never explain how practical reason "is nonetheless supposed to become constitutive again, to give birth out of itself and give itself content." Hegel, *Faith and Knowledge*, trans. Walter Cerf and H. S. Harris (Albany: SUNY Press, 1977), 80.

283. Whether Kant appreciated the radicality of his argument, a strategy that began with the epigenesis of concepts but gradually became so encompassing that even morality grew out of an epigenetic reason, is not clear. It has, however, caused some of Kant's interpreters to draw back in the face of this portrait of Kant. As Hans Ingensiep put it, for example, if Kant were to have taken the epigenesis of the categories seriously, then "so müßte er im letzten doch wieder ins übersinnliche Substrat der Natur verweisen.... Die Kategorien gehören zur intelligiblen Welt; sie wären mit Hypothesen zu einem phänomenalen epigenestischen Ursprung nicht faßbar." "Die biologischen Analogien und die erkenntnistheoretischen Alternativen," 393. Ingensiep is mistaken, however, in thinking that this suggests a consequence that would be out of line with Kant's approach to reason. That is, within Kant's system, reason does effectively function as a "supersensible substrate" so far as it understands its relationship to the realms of both nature and freedom. Kant was consistent, moreover, in rejecting positive discussions of epigenesis as a biological phenomenon in nature, even as he repeatedly appealed to this as the model for understanding the *metaphysical* generation of reason and the categories alike. In light of this it cannot be right to suggest, as a number of Kant's interpreters have done, that Kant restricted his understanding of epigenesis to the epigenesis of experience. This conclusion is reached by mapping Kant's later hypothesis regarding a "generic preformation" of the species lines (5:423) onto the production of experience, such that the categories perform their role on the generically preformed side of the equation, while their construction of experience results in the active generation of something new. For this line of interpretation see Haffner, "Die Epigenesisanalogie in Kants Kritik" (1997); Ingensiep, "Die biologischen Analogien und die erkenntnistheoretischen Alternativen" (1994); Ingensiep, "Organism, Epigenesis, and Life" (2006); Piché, "The Precritical Use of the Metaphor of Epigenesis" (2001); and Shaw, "Function and Epigenesis in Kant's *Critique of Pure Reason*."(2003). The analogy between generic preformation and the construction of experience might hold in a limited sense (assuming the caveat that the categories, while lawlike, are not preformed). But this is not what Kant meant by the "epigenesis of reason," since the very basis of Kant's long-standing

attraction to epigenesis was its ability to position the mind's independence from both sense and God as suppliers of mental form. Kant was a *metaphysician*, not a naturalist, with respect to reason, and because of this he was attracted to a metaphysical conception of reason as something self-born, even as he remained suspicious of the models of emergent vital forces that were being proposed in the life sciences of his day.

284. Kant was clear regarding their identity: "practical reason has the same cognitive faculty for its foundation as the speculative, so far as they are both pure reason" (5:90; cf. 6:382). But he was also delighted by the manner in which their investigation had proceeded in identical ways. As he summarized his findings in the analytic of practical reason, "Here I wish to call attention, if I may, to one thing, namely, that every step which one takes with pure reason, even in the practical field where one does not take subtle speculation into account, so neatly and naturally dovetails with all parts of the *Critique of Pure* (theoretical) *Reason* that it is as if each step had been carefully thought out merely to establish this connection" (5:106). It was precisely because of this that Kant felt confident in pursuing the strategy he had followed in the first *Critique* with respect to identifying the table of judgments as the genealogical basis of both the categories and the ideas of reason; in this case, with respect to the genetic grounds upon which he could identify causality and freedom (5:55–57, 5:65–67, 5:68–70).

285. Timothy Lenoir appeals to a similar distinction when distinguishing between a mechanical series of causes and effects as a "linear series" and the teleological approach to causality as a "reflexive series" according to which the end of the series is simultaneously the cause of it. Lenoir discusses this in terms of what he takes to be the "teleomechanism" of Kant's approach to organic life in the *Critique of Judgment* (e.g., 5:373). See Lenoir, *The Strategy of Life: Teleology and Mechanics in Nineteenth-Century German Biology* (Chicago: University of Chicago Press, 1989), 25.

286. Günter Zöller regards the tracks between Kant's critical doctrines and his anthropological works to be necessarily parallel rather than entwined, and as describing therefore only a "mutually supplementary relation" between the critical theory of reason and the natural history of reason. "Kant's Political Anthropology," *Kant Jahrbuch* 3 (2011): 131–161. Paul Menzer argues similarly regarding the need to keep Kant's projects distinct in *Kants Lehre von der Entwicklung in Natur und Geschichte* (Berlin: G. Reimer, 1911), 404–445. Since I take it that Kant's use of organic models has a deep methodological impact on the critical system—the "epigenesis of reason" does not only have a *metaphorical* value for Kant, in other words—I am willing to reach the stronger conclusion regarding the necessary intertwining of Kant's critical and anthropological concerns regarding reason's historical development.

287. This distinction between claims made about reason in contrast to those concerning nature does more than simply demonstrate Kant's lifelong at-

tention to the specter of subreption; it locates him among the vanguard of those concerned with the establishment of scientific practices regarding "boundary maintenance." In the mid-1770s Lavoisier ended the phlogiston debate in large part because his oxygen theory offered a new vocabulary, a new method, and above all, a severely circumscribed set of questions that the chemical scientist could ask. The model would be adopted by geology in the coming decades, and indeed all of the sciences established in the nineteenth century would eventually follow suit, with the key to their successful establishment in each case being determined by such boundary maintenance. Questions regarding (or coming out of) metaphysical speculation, religious presuppositions, or biblical interpretation—in short, questions relying on claims that were untestable and therefore unknowable—would simply lie outside of the boundaries of a given science. This is perhaps why one might say that Kant's German idealist successors were mistaken to have ignored Kant's boundaries when establishing their own systems, even if they got Kant right in taking his organic conception of reason as their starting point.

288. Kant's conclusions regarding the relationship between these modes of judgment developed directly out of his discussions of physicotheology and moral teleology in the first *Critique*, even if these were to be freshly distinguished in order demonstrate that only moral teleology was capable of yielding conviction in its proofs (5:462, 5:478).

289. Although much has been made of Kant's endorsement of Blumenbach and of questions regarding Blumenbach's influence on Kant in his discussion of epigenesis, one should not forget that, whatever influence might be claimed, Blumenbach in fact transgressed a clear boundary set by Kant between thinking about nature as purposive and claiming that nature was in fact purposive. Robert J. Richards emphasizes this difference between Kant and Blumenbach in "Kant and Blumenbach on the *Bildungstrieb*: A Historical Misunderstanding," *Studies in History and Philosophy of Biology and Biomedical Science* 31 (2000): 11–32. See also Richards's *The Romantic Conception of Life: Science and Philosophy in the Age of Goethe* (Chicago: University of Chicago Press, 2002), chap. 5., pp. 216–237. As Timothy Lenoir describes Blumenbach's position, "The *Bildungstrieb* was not a blind mechanical force of expansion which produced structure by being opposed in some way; it was not a chemical force of 'fermentation,' nor was it a soul superimposed on matter. Rather the *Bildungstrieb* was conceived as a teleological agent which had its antecedents ultimately in the inorganic realm but which was an emergent vital force." "Kant, Blumenbach, and Vital Materialism," 83. It was precisely this interpenetration of form and force—something Kant explicitly liked about Blumenbach's theory—that caused Caspar Wolff, the first author to describe vegetative growth and reproduction as a form of epigenesis, to complain about Blumenbach's position. For Wolff, force simply could not by definition also be responsible for

form. See Wolff, "Von der eigenthümlichen und wesentlichen Kraft der vegetabilischen sowohl als auch der animalischen Substanz," in *Zwo Abhandlungen über die Nutritionskraft welche von der Kayserlichen Akademie der Wissenschaft in St. Petersburg den Preis getheilt haben*. St. Petersburg: Kayserliche Akademie der Wissenschaften, 1789.
290. Kant's caution regarding the progress of the life sciences has continued relevance today. After nearly a century dominated by the genes-as-destiny model, the resistance of the organism to this kind of determination has formed the core of a recent reorganization in genetic investigations, a reframing made necessary by the discovery of the central role played by emergent, environmentally fluid switches for gene expression. The new science surrounding this discovery is called "epigenetics."

EPILOGUE

291. Kant had played with this sort of image as early as 1772, imagining as well a difference in the inhabitants disposed to living in one or the other of the various "regions" of reason. As Kant pictured this geography in 1772, however, it was dogmatic metaphysics that formed an island of cognition, and bridges between this island and the mainland of experience were still thought to be possible: "In metaphysics, like an unknown land of which we intend to take possession, we have first assiduously investigated its situation and access to it. (It lies in the (region) hemisphere of pure reason;) we have even drawn the outline of where this island of cognition is connected by bridges to the land of experience, and where it is separated by a deep sea; we have even drawn its outline and are as it were acquainted with its geography (ichnography), be we do not know what might be found in this land, which is maintained as uninhabitable by some people and to be their real domicile by others. We will take the general history of this land of reason into account in accordance with this general geography" (17:559).
292. Goethe, "Judgment through Intuitive Perception" (1817), in *Scientific Studies*, trans. Douglas Miller (New York: Suhrkamp, 1988), 12:32. I discuss Kant in relationship to Goethe on this point more fully in "Intuition and Nature in Kant and Goethe."
293. As Darwin put it, "This resemblance is often expressed by the term 'unity of type': or by saying that the several parts and organs in the different species of the class are homologous. The whole subject is included under the general name of Morphology. This is the most interesting department of natural history, and may be said to be its very soul." *The Origin of Species*, 415.

Bibliography

Abrams, M. H. *The Mirror and the Lamp: Romantic Theory and the Critical Tradition.* New York: Norton, 1958.
Acta Eruditorum. "Ouvràges divers de Mr. de Maupertuis. Amsterdam, 1744." Unsigned review of 1744 edition of *Oeuvres*, by P. L. M. Maupertuis. 1745, 221–229.
Allan, D. G. C., and R. E. Schofield. *Stephen Hales, Scientist and Philanthropist.* London: Scolar, 1980.
Allison, H. E. *The Kant-Eberhard Controversy.* Baltimore: Johns Hopkins University Press, 1973.
———. "Transcendental Affinity—Kant's Answer to Hume." In *Kant's Theory of Knowledge: Selected Papers from the Third International Kant Congress*, ed. L. W. Beck, 119–127. Dordrecht: Reidel, 1974.
———. *Kant's Theory of Freedom.* Cambridge: Cambridge University Press, 1990.
———. *Kant's Transcendental Idealism.* 2nd ed. New Haven, CT: Yale University Press, 2004.
Ameriks, K., and S. Naragon. Introduction to *Lectures on Metaphysics*, by Immanuel Kant. Trans. and ed. K. Ameriks and S. Naragon. Cambridge: Cambridge University Press, 1997.
———. *Kant and the Fate of Autonomy: Problems in the Appropriation of the Critical Philosophy.* Cambridge: Cambridge University Press, 2000.
———. *Kant's Theory of Mind: An Analysis of the Paralogisms of Pure Reason.* 2nd ed. Oxford: Oxford University Press, 2000.
Anstey, P. *The Philosophy of Robert Boyle.* New York: Routledge, 2000.
———. "Boyle on Seminal Principles." *Studies in the History and Philosophy of Biological and Biomedical Sciences* 33 (2002): 597–630.

———. "Robert Boyle and Locke's 'Morbus' Entry: A Reply to J. C. Walmsley." *Early Science and Medicine* 7 (2002): 358–377.
Anstey, P., and S. Harris. "Locke and Botany." *Studies in the History and Philosophy of Biological and Biomedical Science* 37 (2006): 151–171.
Aristotle. *Generation of Animals*. Trans. A. L. Peck. Cambridge, MA: Harvard University Press, 1963.
———. *De Anima*. Trans. J. A. Smith. In *The Complete Works of Aristotle*, ed. J. Barnes, vol. 1. Princeton, NJ: Princeton University Press, 1984.
———. *Parts of Animals*. Trans. W. Ogle. In *The Complete Works of Aristotle*, ed. J. Barnes, vol. 1. Princeton, NJ: Princeton University Press, 1984.
Arthur, R. "Animal Generation and Substance in Sennert and Leibniz." In *The Problem of Animal Generation in Early Modern Philosophy*, ed. J. E. H. Smith, 147–174. Cambridge: Cambridge University Press, 2006.
Aucante, V. "Descartes's Experimental Method and the Generation of Animals." In *The Problem of Animal Generation in Early Modern Philosophy*, ed. J. E. H. Smith, 65–79. Cambridge: Cambridge University Press, 2006.
Baumgarten, A. G. *Metaphysica*. 3rd ed. Halle: C. H. Hemmerde, 1757.
Beck, L. W. *Early German Philosophy: Kant and His Predecessors*. Cambridge, MA: Belknap, 1969.
———. *Essays on Kant and Hume*. New Haven, CT: Yale University Press, 1978.
———. "Lambert and Hume in Kant's Development." In *Essays on Kant and Hume*. New Haven, CT: Yale University Press, 1978.
Beguelin, N. "Conciliation des idées de Newton et de Leibnitz sur l'espace et le vuide [sic]." 1769. *Histoire de l'Académie Royale des Sciences et des Belles-Lettres de Berlin*. Berlin: Haude et Spenner, 1771.
Beiser, F. C. *The Fate of Reason: German Philosophy from Kant to Fichte*. Cambridge, MA: Harvard University Press, 1987.
Bernasconi, R. "Who Invented the Concept of Race? Kant's Role in the Enlightenment Construction of Race." In *Race*, ed. R. Bernasconi, 11–36. Oxford: Blackwell, 2001.
———. "Kant's Third Thoughts on Race." In *Reading Kant's Geography*, ed. S. Elden and E. Mendietta, 291–318. Albany: SUNY Press, 2011.
Bernier, F. "Nouvelle division de la terre, par les différentes espèces ou races d'hommes qui l'habitent." *Journal des sçavans* (1684): 133–140. Reprint (1685): 148–155.
Birken-Bertsch, H. *Subreption und Dialektik bei Kant: Der Begriff des Fehlers der Erschleichung in der Philosophie des 18. Jahrhunderts*. Stuttgart: Frommann-Holzboog, 2006.
Blair, P. *Botanick Essays in Two Parts*. London: W. and J. Innys, 1720.
———. "Observations Upon the Generation of Plants, in a Letter to Hans Sloan." *Philosophical Transactions* 31 (1720–1721): 216–221.
Blumenbach, J. *Über den Bildungstrieb*. Göttingen: Dietrich, 1781.
———. *On the Natural Varieties of Mankind*. 1775. 3rd ed., 1795. Trans. T. Bendyshe, 1865. Reprint, New York: Bergman, 1969.

Bonnet, C. *Contemplation de la nature*. 2 vols. Amsterdam: M.-M. Rey, 1764–1766.
———. *Betrachtung über die Natur*. Trans. J. D. Titius. 2 vols. Leipzig: J. F. Junius, 1766; 2nd ed., 1772.
———. *La palingénésie philosophique*. 2 vols. Geneva: C. L. Philibert and B. Chirol, 1770.
———. *Philosophische Palingenesie oder Gedanken über den vergangenen und künftigen Zustanden lebender Wesen*. 2nd ed. Trans. J. C. Lavater. 2 vols. Zurich: Orell, Gessner, Fuesili, 1770.
———. *Oeuvres d'histoire naturelle et de philosophie*. 10 vols. Neuchatel: S. Fauche, 1779–1783.
Bortoft, H. *The Wholeness of Nature: Goethe's Way of Science*. Edinburgh: Floris Books, 1996.
Bourguet, L. *Lettres philosophiques sur la formation des sels et des crystaux et sur la génération et le méchanisme organique des plantes et des animaux*. Amsterdam: F. L'Honoré, 1729.
Bowler, P. J. "Preexistence and Preformation in the Seventeenth Century: A Brief Analysis." *Journal of the History of Biology* 4 (1971): 221–244.
Boyle, R. "Of the Imperfection of the Chemists' Doctrine of Qualities." In *Selected Philosophical Papers of Robert Boyle*, ed. M. A. Stewart. Indianapolis: Hackett, 1991.
———. "The Origin of Forms and Qualities According to the Corpuscular Philosophy." In *Selected Philosophical Papers of Robert Boyle*, ed. M. A. Stewart. Indianapolis: Hackett, 1991.
———. *Selected Philosophical Papers of Robert Boyle*. Ed. M. A. Stewart. Indianapolis: Hackett, 1991.
———. "Considerations and Experiments, Touching the Origin of Qualities and Forms. The Historical Part." In *The Works of Robert Boyle*, ed. M. Hunter and E. B. Davis, vol. 5. London: Pickering and Chatto, 1999.
———. *The Works of Robert Boyle*. Ed. M. Hunter and E. B. Davis. 14 vols. London: Pickering and Chatto, 1999.
Brandt, R. "Materialien zur Entstehung der *Kritik der reinen Vernunft* (John Locke und Johann Schultz)." In *Beiträge zur Kritik der reinen Vernunft, 1781–1981*, ed. I. Heidemann and W. Ritzel, 37–68. Berlin: De Gruyter, 1981.
Buffon, G.-L. "Préface du traducteur." In *La statique des végétaux, et l'analyse de l'air*, by Stephen Hales. Paris: J. Vincent, 1735.
———. Preface to *La methode des fluxions, et des suites infinies*, by I. Newton. Paris: Debure l'aîné, 1740.
———. "Initial Discourse." In *Histoire naturelle, générale et particulière*, vol. 1. Paris: Imprimerie Royale, 1749.
———. *Histoire naturelle, générale et particulière*. 15 vols. Paris: Imprimerie Royale, 1749–1767.
———. *Allgemeine Historie der Natur: Nach ihren besonderen Theilen abgehandelt*. Trans. A. G. Kästner. Hamburg: G. C. Grund and A. H. Holle, 1750.

———. *Natural History, General and Particular*. Trans. W. Smellie. 9 vols. Edinburgh: W. Creech, 1780.

———. "Of Nutrition and Growth." Chap. 3 of *History of Animals*. 1749. In *Natural History, General and Particular*, trans. W. Smellie, vol. 2. Edinburgh: W. Creech, 1780.

———. "Of the Degeneration of Animals." 1766. In *Natural History, General and Particular*, trans. W. Smellie, vol. 7. Edinburgh: W. Creech, 1780.

———. "Of the Expansion, Growth, and Delivery of the Foetus." Chap. 11 of *History of Animals*. 1749. In *Natural History, General and Particular*, trans. W. Smellie, vol. 2. Edinburgh: W. Creech, 1780.

———. "Second View of Nature." 1765. In *Natural History, General and Particular*, trans. W. Smellie, vol. 7. Edinburgh: W. Creech, 1780.

———. *Universal Natural History*. In *Natural History, General and Particular*, trans. W. Smellie, vol. 1. Edinburgh: W. Creech, 1780–1785.

———. "The Natural History of Man." Sect. 9, "Of the Varieties of the Human Species." 1749. In *Natural History, General and Particular*, 3rd ed., trans. W. Smellie, vol. 3. London: A. Strahan and T. Cadell, 1791.

———. "On Degeneration." 1766. In *Natural History, General and Particular*, 3rd ed., trans. W. Smellie, vol. 7. London: A. Strahan and T. Cadell, 1791.

———. *Buffon's Natural History*. Trans. J. S. Barr. 10 vols. London: H. D. Symonds, 1797.

———. "A Comparison Between Animals, Vegetables, and Other Productions of Nature." Chap. 1 of *History of Animals*. In *Buffon's Natural History*, trans. J. S. Barr, vol. 2. London: H. D. Symonds, 1797.

———. "Experiments on the Method of Generation." Chap. 6 of *History of Animals*. In *Buffon's Natural History*, trans. J. S. Barr, vol. 3. London: H. D. Symonds, 1797.

———. "Of Infancy." Chap. 2 of *History of Man*. In *Buffon's Natural History*, trans. J. S. Barr, vol. 3. London: H. D. Symonds, 1797.

———. "Of Nutrition and Growth." Chap. 3 of *History of Animals*. In *Buffon's Natural History*, trans. J. S. Barr, vol. 2. London: H. D. Symonds, 1797.

———. "Of Reproduction in General." Chap. 2 of *History of Animals*. In *Buffon's Natural History*, trans. J. S. Barr, vol. 2. London: H. D. Symonds, 1797.

———. "Of the Expansion, Growth, and Delivery of the Foetus." Chap. 11 of *History of Animals*. In *Buffon's Natural History*, Trans. J. S. Barr, vol. 3. London: H. D. Symonds, 1797.

———. *Oeuvres philosophiques de Buffon*. Ed. J. Piveteau. Paris: Presses Universitaires de France, 1954.

———. "The 'Initial Discourse' to Buffon's *Histoire naturelle*: The First Complete English Translation." Trans. J. Lyon. *Journal of the History of Biology* 9 (1976): 133–181.

———. "Initial Discourse." In *From Natural History to the History of Nature: Readings from Buffon and His Critics*, trans. P. Sloan and J. Lyon, 89–128. Notre Dame, IN: University of Notre Dame Press, 1981.

———. "Préface du traducteur." In *From Natural History to the History of Nature: Readings from Buffon and His Critics*, trans. P. Sloan and J. Lyon, 35–40. Notre Dame, IN: University of Notre Dame Press, 1981.

Buroker, J. V. *Space and Incongruence: The Origin of Kant's Idealism*. Dordrect: D. Reidel, 1980.

Calvin, J. *Institutes of the Christian Religion*. Ed. J. McNeil. 2 vols. Philadelphia: Westminster, 1960.

Camerarius, R. J. Letter to Michael Valentin under the title *Academiae Caesareo Leopold. N.C. Hecotorus II. Rudolphi Jacobi Camerarii, Professoris Tubingensis, ad Thessalum, D. Mich. Bernardum Valentini Professorum Giessensem excellentissimum, de sexu plantarum epistola*. Tubingae: Typis Viduae Rommeii, 1694.

———. *Ueber das Geschlecht der Pflanzen (De sexu plantarium epistola)*. 1694. Ed. and trans. M. Möbius. Leipzig: W. Engelmann, 1899.

Carl, W. "Kant's First Drafts of the Deduction of the Categories." In *Kant's Transcendental Deductions: The Three Critiques and the Opus Postumum*, ed. E. Förster, 3–20. Stanford, CA: Stanford University Press, 1989.

———. *Der schweigende Kant: Die Entwürfe zu einer Deduktion der Kategorien vor 1781*. Göttingen: Vandenhoeck and Ruprecht, 1989.

Carroll, S. B. *Endless Forms Most Beautiful: The New Science of Evo Devo and the Making of the Animal Kingdom*. New York: W. W. Norton, 2005.

Castillon, J. de. "Descartes et Locke conciliés." In *Nouveaux Memoires de l'Académie Royale des Sciences et Belles-Lettres*. Berlin: Voss, 1772.

Cesalpino, A. *De plantis libri XVI*. Bk. I. Florence, 1583.

Clericuzio, A. *Elements, Principles, and Corpuscles: A Study of Atomism and Chemistry in the Seventeenth Century*. Dordrecht: Kluwer, 2000.

Code, A. "Soul as Efficient Cause in Aristotle's Embryology." *Philosophical Topics* 15 (1986): 51–60.

Cole, F. J. "Microscopic Science in Holland in the Seventeenth Century." *Journal of the Quekett Microscopical Club* 4 (1938): 59–77.

Cooper, J. M. "Metaphysics in Aristotle's Embryology." *Proceedings of the Cambridge Philological Society* 214 (1988): 14–41.

Darwin, C. *The Autobiography of Charles Darwin, 1809–1882, with Original Omissions Restored*. Ed. N. Barlow. London: Collins, 1958.

———. *The Origin of Species by Means of Natural Selection; or, The Preservation of Favoured Races in the Struggle for Life*. 1859. Ed. J. W. Burrow. London: Penguin Books, 1968.

Deason, G. B. "Reformation Theology and the Mechanistic Conception of Nature." In *God and Nature: Historical Essays on the Encounter between Christianity and Science*, ed. D. C. Lindberg and R. L. Numbers, 167–191. Berkeley: University of California Press, 1986.

De Risi, V. *Geometry and Monadology: Leibniz's Analysis Situs and Philosophy of Space*. Basel: Birkhäuser, 2007.
Descartes, Rene. *La description du corps humain*. 1664. In *Oeuvres de Descartes*, ed. C. Adam and P. Tannery, 11:223–257. Paris: Vrin, 1964–1976.
———. Descartes, Rene. *L'homme de René Descartes*. 1664. In *Oeuvres de Descartes*, ed. C. Adam and P. Tannery, 11:119–202. Paris: Vrin, 1964–1976.
———. *Description of the Human Body*. In *The Philosophical Writings of Descartes*, trans. J. Cottingham, R. Stoothoff, and D. Murdoch, 1:314–324. Cambridge: Cambridge University Press, 1985.
———. *Treatise on Man*. In *The Philosophical Writings of Descartes*, trans. J. Cottingham, R. Stoothoff, and D. Murdoch, 1:79–108. Cambridge: Cambridge University Press, 1985.
Des Chene, D. "From Natural Philosophy to Natural Science." In *The Cambridge Companion to Early Modern Philosophy*, ed. D. Rutherford, 67–94. Cambridge: Cambridge University Press, 2006.
Detlefsen, K. "Explanation and Demonstration in the Haller-Wolff Debate." In *The Problem of Animal Generation in Early Modern Philosophy*, ed. J. E. H. Smith, 235–261. Cambridge: Cambridge University Press, 2006.
De Vleeschauwer, H. J. *La déduction transcendentale dans l'oeuvre de Kant*. 2 vols. Paris: É. Champion, 1934.
———. *The Development of Kantian Thought: The History of a Doctrine*. Trans. A. R. C. Duncan. London: T. Nelson and Sons, 1962.
Dörflinger, B. *Das Leben theoretischer Vernunft*. Berlin: W. de Gruyter, 2000.
Downing, L. "Locke's Ontology." In *The Cambridge Companion to Locke's "Essay Concerning Human Understanding,"* ed. L. Newman, 352–380. Cambridge: Cambridge University Press, 2007.
Duchesneau, F. "Épigenèse de la raison pure et analogies biologiques." In *Kant Actuel: Homage à Pierre Laberge*, ed. F. Duchesneau, G. Lafrance, and C. Piché, 233–256. Montreal: Bellarmine, 2000.
———. "Louis Bourguet et le modèle des corps organiques." In *Antonio Vallisneri: L'édizione del testo scientifico d'età moderna*, ed. M. T. Monti, 125–164. Florence: L. Olschki, 2003.
———. "Charles Bonnet's Neo-Leibnizian Theory of Organic Bodies." In *The Problem of Animal Generation in Early Modern Philosophy*, ed. J. E. H. Smith, 285–314. Cambridge: Cambridge University Press, 2006.
———. "'Essential Force' and 'Formative Force': Models for Epigenesis in the Eighteenth Century." In *Self-Organization and Emergence in Life Sciences*, ed. B. Feltz, M. Crommelinck, and P. Goujon, 171–186. Dordrecht: Springer, 2006.
Engel, J. J. *Der Philosoph für die Welt*. Part 2. Leipzig, 1777.
Farley, J. *The Spontaneous Generation Controversy from Descartes to Oparin*. Baltimore: Johns Hopkins University Press, 1977.
Fischer, K. P. "John Locke in the German Enlightenment." *Journal of the History of Ideas* 36, no. 3 (1975): 431–446.

Fisher, M. "Kant's Explanatory Natural History: Generation and Classification of Organisms in Kant's Natural Philosophy." In *Understanding Purpose: Kant and the Philosophy of Biology*, ed. P. Huneman, 101–121. North American Kant Society Studies in Philosophy, vol. 8. Rochester, NY: University of Rochester Press, 2007.

———. "Organisms and Teleology in Kant's Natural Philosophy." Ph.D. diss., Emory University, 2008.

Förster, E. "Die Bedeutung von §§76, 77 der Kritik der Urteilskraft für die Entwicklung der nachkantischen Philosophie." *Zeitschrift für philosophische Forschung* 56 (2002): 169–190.

———. *Die 25 Jahre der Philosophie*. Frankfurt: Klostermann, 2011.

Friedman, M. *Kant and the Exact Sciences*. Cambridge, MA: Harvard University Press, 1992.

Gadamer, H.-G. *Truth and Method*. Trans. J. Weinsheimer and D. Marshall. New York: Continuum, 2004.

Garber, D. "Leibniz: Physics and Philosophy." In *The Cambridge Companion to Leibniz*, ed. N. Jolley, 270–352. Cambridge: Cambridge University Press, 1995.

Garber, D., and B. Longuenesse, eds. *Kant and the Early Moderns*. Princeton, NJ: Princeton University Press, 2008.

Gasking, E. *Investigations into Generation, 1651–1828*. London: Hutchinson, 1967.

Genova, A. C. "Kant's Epigenesis of Pure Reason." *Kant-Studien* 65, no. 3 (1974): 259–273.

Gigante, D. *Life: Organic Form and Romanticism*. New Haven, CT: Yale University Press, 2009.

Ginsborg, H. "Kant on Understanding Organisms as Natural Purposes." In *Kant and the Sciences*, ed. E. Watkins, 231–258. Oxford: Oxford University Press, 2001.

Glass, B. "Heredity and Variation in the Eighteenth Century Concept of the Species." In *Forerunners of Darwin, 1745–1859*, ed. B. Glass, O. Temkin, and W. L. Strauss, 144–172. Baltimore: Johns Hopkins Press, 1968.

Goethe, J. W. von. "Later Studies and Collections." In *Goethe's Botanical Writings*, ed. and trans. B. Mueller. Woodbridge, CT: Ox Bow, 1952.

———. "On the Spiral Tendency in Plants." In *Goethe's Botanical Writings*, ed. and trans. Bertha Mueller. Woodbridge, CT: Oxbow, 1952.

———. "Fortunate Encounter." In *Goethe: Scientific Studies*, vol. 12 of *Goethe's Collected Works*, ed. and trans. D. Miller. New York: Suhrkamp, 1983.

———. "Leaf and Root." In *Goethe: Scientific Studies*, trans. D. Miller. New York: Suhrkamp, 1983.

———. "Blatt und Wurzel." In *Goethes Schriften zur Morphologie*, ed. D. Kuhn. Frankfurt: Deutscher Klassiker, 1988.

———. "Nacharbeiten und Sammlungen." In *Goethes Schriften zur Morphologie*, ed. D. Kuhn. Frankfurt: Deutscher Klassiker, 1988.

———. "Later Studies and Collections." In *Goethe's Botanical Writings*, ed. and trans. B. Mueller. Woodbridge, CT: Oxbow, 1989.

———. "The Metamorphosis of Plants." Trans. H. Norden. In *The Metamorphosis of Plants*, introduction and photography by G. L. Miller. Cambridge, MA: MIT Press, 2009.

Gough, J. W. "John Locke's Herbarium." *Bodleian Library Record* 7 (1962–1967): 42–46.

Grier, M. *Kant's Doctrine of Transcendental Illusion*. Cambridge: Cambridge University Press, 2001.

Guerlac, H. *Lavoisier—The Crucial Year: The Background and Origin of His First Experiments on Combustion in 1772*. Ithaca, NY: Cornell University Press, 1961.

———. "Hales, Stephen." In *Dictionary of Scientific Biography*, ed. C. Gillispie, 35–48. New York: Charles Scribner's Sons, 1970–1980.

———. *Essays and Papers in the History of Modern Science*. Baltimore: Johns Hopkins University Press, 1977.

Guyer, P. *Kant and the Claims of Knowledge*. Cambridge: Cambridge University Press, 1987.

———. "Organisms and the Unity of Science." In *Kant and the Sciences*, ed. E. Watkins, 259–281. Oxford: Oxford University Press, 2001.

———. *Kant's System of Nature and Freedom: Selected Essays*. Oxford: Oxford University Press, 2005.

Haffner, T. "Die Epigenesisanalogie in Kants Kritik der reinen Vernunft." Ph.D. diss., Universität des Saarlandes, 1997.

Hales, S. *La statique des végétaux, et l'analyse de l'air*. Trans. G.-L. Buffon. Paris: Debure, 1735.

———. *Statick der Gewächse oder angestelte Versuche mit dem Saft in Pflanzen und ihren Wachstum*. Trans. C. Wolff. Halle: Rengerischen Buchhandlung, 1748.

———. *Vegetable Staticks*. 1727. Reprint, London: Scientific Book Guild, 1961.

Hall, T. S. "Biological Analogs of Newtonian Paradigms." *Philosophy of Science* 35 (1968): 6–27.

Haller, A. von. *The Natural Philosophy of Albrecht von Haller*. Ed. and trans. S. Roe. New York: Arno, 1981.

———. "Reflections on the Theory of Generation of Mr. Buffon." In *From Natural History to the History of Nature: Readings from Buffon and His Critics*, ed. and trans. P. Sloan and J. Lyon, 314–327. Notre Dame, IN: University of Notre Dame Press, 1981.

Hankins, T. L. *Science and the Enlightenment*. Cambridge: Cambridge University Press, 1985.

Hartnack, A. *Geschichte der Königlich Preussischen Akademie der Wissenschaften zu Berlin*. 4 vols. in 3. Berlin: Reichsdruckerei, 1900.

Harvey, W. *Disputations Touching the Generation of Animals*. 1651. Trans. G. Witteridge. Reprint, Oxford: Blackwell Scientific, 1981.

Herder, J. *Sämtliche Werke*. Ed. B. Suphan. 33 vols. Berlin: Weidmann, 1883.
Hegel, G. W. F. *Faith and Knowledge*. Trans. W. Cerf and H. S. Harris. Albany: SUNY Press, 1977.
Henrich, D. "The Proof-Structure of Kant's Transcendental Deductions." *Review of Metaphysics* 22 (1969): 640–659.
———. "Kant's Notion of a Deduction and the Methodological Background of the First *Critique*." In *Kant's Transcendental Deductions*, ed. E. Förster, 29–46, 251–252. Stanford, CA: Stanford University Press, 1989.
Henry, J. "Occult Qualities and the Experimental Philosophy: Active Principles in Pre-Newtonian Matter Theory." *History of Science* 24 (1986): 335–381.
———. "Themistius and Spontaneous Generation in Aristotle's Metaphysics." *Oxford Studies in Ancient Philosophy* 24 (2003): 183–208.
———. "Understanding Aristotle's Reproductive Hylomorphism." *Apeiron: A Journal of Ancient Philosophy and Science* 39 (2006): 269–300.
Heringman, N., ed. *Romantic Science: The Literary Forms of Natural History*. Albany: SUNY Press, 2003.
Herz, M. *Betrachtungen aus der spekulativen Weltweisheit*. Hamburg: F. Meiner, 1990.
Hume, D. *A Treatise of Human Nature*. Ed. L. A. Selby-Bigge. Rev. ed., ed. P. H. Nidditch. Oxford: Oxford University Press, 1978.
Huneman, P. "Naturalising Purpose: From Comparative Anatomy to the 'Adventure of Reason.'" *Studies in the History and Philosophy of Biological and Biomedical Science* 37 (2006): 649–674.
———. *Métaphysique et biologie: Kant et la constitution du concept d'organisme*. Paris: Éditions Kimé, 2008.
Ingensiep, H. W. "Die Biologischen Analogien und die erkenntnistheoretischen Alternativen in Kant's Kritik der reinen Vernunft B §27." *Kant-Studien* 85, no. 4 (1994): 381–393.
———. "Organism, Epigenesis, and Life in Kant's Thinking." *Annals of the History and Philosophy of Biology* 11 (2006): 59–84.
Jacquette, D., ed. *Philosophy, Psychology, and Psychologism: Critical and Historical Readings on the Psychological Turn in Philosophy*. Dordrecht: Kluwer Academic, 2003.
Jolley, N., ed. *The Cambridge Companion to Leibniz*. Cambridge: Cambridge University Press, 1995.
Kant, I. *Immanuel Kants Menschenkunde: Oder Philosophische Anthropologie nach handschriftlichen Vorlesungen*. Ed. F. C. Starke. Leipzig, 1831.
———. *Kants Anweisung zur Menschen- und Welterkenntnis*. Ed. F. C. Starke. Leipzig, 1831.
———. *Allgemeine Naturgeschichte und Theorie des Himmels* [Universal natural history and theory of the heavens]. 1755. In *Kants gesammelte Schriften*, vol. 1, ed. J. Rahts and K. Lasswitz. Berlin: W. de Gruyter, 1902.
———. *Die Frage, ob die Erde veralte, physikalisch erwogen* [The question whether the Earth is aging, considered from the point of view of physics]. 1754. In

BIBLIOGRAPHY

Kants gesammelte Schriften, vol. 1, ed. J. Rahts and K. Lasswitz. Berlin: W. de Gruyter, 1902.

———. *Geschichte und Naturbeschreibung der merkwürdigsten Vorfälle des Erdbebens, welches an dem Ende des 1755sten Jahres einen großen Theil der Erde erschüttert hat.* [History and natural description of the most curious occurrences associated with the [Lisbon] earthquake]. 1756. In *Kants gesammelte Schriften*, vol. 1, ed. J. Rahts and K. Lasswitz. Berlin: W. de Gruyter, 1902.

———. "Untersuchung der Frage, ob die Erde in ihrer Umdrehung um die Achse . . ." [Investigation of the question whether the earth in its axial rotation, whereby it causes the change of day and night, has experienced any change since the earliest times of its origin, and how one could answer this question, announced for the current year's prize, by the Royal Academy of Sciences in Berlin]. 1754. In *Kants gesammelte Schriften*, vol. 1, ed. J. Rahts and K. Lasswitz. Berlin: W. de Gruyter, 1902.

———. *Kants gesammelte Schriften*. 29 vols. Berlin: W. de Gruyter, 1902–.

———. *Reflexionen zur Medizin* [Reflections on medicine]. In *Kants gesammelte Schriften*, vol. 15, ed. E. Adickes. Berlin: W. de Gruyter, 1913.

———. *Kants handschriftlicher Nachlass: Reflexionen zur Metaphysik.* In *Kants gesammelte Schriften*, vols. 17–18, ed. E. Adickes. Berlin: W. de Gruyter, 1926–1928.

———. *Critique of Practical Reason.* Trans. L. W. Beck. New York: Macmillan, 1956.

———. *Groundwork of the Metaphysic of Morals.* Trans. H. J. Paton. New York: Harper and Row, 1964.

———. *Critique of Judgment.* Trans. W. Pluhar. Indianapolis: Hackett, 1987.

———. *Concerning the Ultimate Ground of the Different Regions in Space.* In *Kant's Latin Writings: Translations, Commentaries, and Notes*, trans. L. W. Beck. New York: P. Lang, 1992.

———. *Dreams of a Spirit-Seer Elucidated by Dreams of Metaphysics.* In *Immanuel Kant: Theoretical Philosophy, 1755–1770*, trans. D. Walford. Cambridge: Cambridge University Press, 1992.

———. *Immanuel Kant: Theoretical Philosophy, 1755–1770.* Trans. D. Walford. Cambridge: Cambridge University Press, 1992.

———. *Inaugural Dissertation.* In *Kant's Latin Writings: Translations, Commentaries, and Notes*, trans. L. W. Beck. New York: P. Lang, 1992.

———. *Kant's Latin Writings: Translations, Commentaries, and Notes.* Trans. L. W. Beck. New York: P. Lang, 1992.

———. *Lectures on Logic.* Trans. J. M. Young. New York: Cambridge University Press, 1992.

———. *New Elucidation of the First Principles of Metaphysical Knowledge.* In *Immanuel Kant: Theoretical Philosophy, 1755–1770*, trans. D. Walford. Cambridge: Cambridge University Press, 1992.

———. *The Only Possible Argument in Support of a Demonstration of the Existence of God.* Trans. D. Walford. Cambridge: Cambridge University Press, 1992.
———. *Succinct Exposition of Some Meditations on Fire.* In *Kant's Latin Writings: Translations, Commentaries, and Notes*, trans. L. W. Beck. New York: P. Lang, 1992.
———. *Opus Postumum.* Ed. E. Förster. Trans. E. Förster and M. Rosen. Cambridge: Cambridge University Press, 1993.
———. *Conflict of the Faculties.* In *Religion and Rational Theology*, trans. M. Gregor and R. Anchor. New York: Cambridge University Press, 1996.
———. "What Does It Mean to Orient Oneself in Thinking?" In *Religion and Rational Theology*, trans. Allen Wood. Cambridge: Cambridge University Press, 1996.
———. *Critique of Pure Reason.* Trans. P. Guyer and A. Wood. New York: Cambridge University Press, 1998.
———. *On a Discovery Whereby Any New Critique of Pure Reason Is to Be Made Superfluous by an Older One.* 1790. Trans. H. Allison. In *Theoretical Philosophy after 1781*, ed. H. Allison and P. Heath. Cambridge: Cambridge University Press, 2002.
———. *Immanuel Kant: Notes and Fragments.* Trans. P. Guyer, C. Bowman, and F. Rauscher. Cambridge: Cambridge University Press, 2005.
———. *Critique of Pure Reason.* Trans. N. K. Smith. New York: Palgrave Macmillan, 2007.
———. "Of the Different Races of Human Beings." 1775, rev. ed. 1777. In *The Works of Immanuel Kant: Anthropology, History, and Education*, trans. H. Wilson and G. Zöller, 84–97. Cambridge: Cambridge University Press, 2007.
Kirk, G. S., J. E. Raven, and M. Schofield, trans. and eds. "Anaxagoras of Clazomenae." In *The Presocratic Philosophers*, 2nd ed. Cambridge: Cambridge University Press, 1983.
Kitcher, P. *Kant's Thinker.* Oxford: Oxford University Press, 2011.
Kleingeld, P. "Kant's Second Thoughts on Race." *Philosophical Quarterly* 57 (2007): 573–592.
Krug, W. T. "Bildung." In *Allgemeines Handwörterbuch der philosophischen Wissenschaften nebst ihrer Literatur and Geschichte*, 2nd ed., 1:358–360. Leipzig: F. A. Brockhaus, 1832.
Kruk, R. "A Frothy Bubble: Spontaneous Generation in the Medieval Islamic Tradition." *Journal of Semitic Studies* 35 (1990): 265–282.
Kuehn, M. *Scottish Common Sense in Germany, 1768–1800: A Contribution to the History of Critical Philosophy.* Kingston, ON: McGill-Queen's University Press, 1987.
———. *Kant: A Biography.* Cambridge: Cambridge University Press, 2000.
———. "Kant's Teachers in the Exact Sciences." In *Kant and the Sciences*, ed. E. Watkins, 11–30. Oxford: Oxford University Press, 2001.

Kusch, M. *Psychologism: A Case Study in the Sociology of Philosophical Knowledge.* London: Routledge, 1995.

Lagier, R. *Les races humaines selon Kant.* Paris: Presses Universitaires de France, 2004.

Lambert, J. H. *Neues Organon.* 1764. In *Gesammelte Philosophische Schriften.* Reprint, Hildesheim: Olms, 1965.

Larrimore, M. "Sublime Waste: Kant on the Destiny of the Races." *Canadian Journal of Philosophy* 29 (1999): 99–125.

Laywine, A. *Kant's Early Metaphysics and the Origins of Critical Philosophy.* North American Kant Society Studies in Philosophy, vol. 3. Atascadero, CA: Ridgeview, 1993.

———. Laywine, A. "Kant in Reply to Lambert on the Ancestry of Metaphysical Concepts." *Kantian Review* 5 (2001): 1–48.

Leeuwenhoek, A. van. "Part of a Letter from Mr. Antony van Leeuwenhoek, F.R.S. Concerning Green Weeds Growing in Water, and Some Animalcula Found about Them." *Philosophical Transactions* 23 (1702–1703): 1304–1311.

Leibniz, G. W. "Epistola ad M. Gott. Hanschium, De Enthusiasmo Platonico." July 25, 1707. In *Epistola Godefridi Guilielmi Leibnitii ad Michaelem Gottlieb Hanschium*, ed. G. Veesenmeyer. Leipzig: Gleditsch, 1716.

———. "Epistola ad M. Gott. Hanschium, De Enthusiasmo Platonico." July 25, 1707. In *Gothofredi Guillelmi Leibnitii, Opera Omnia*, ed. L. Dutens, vol. 2. Geneva: Fratre de Tournes, 1768.

———. "Letter to Hansch on Platonic Philosophy or on Platonic Enthusiasm." In *Philosophical Papers and Letters*, ed. and trans. L. Loemker, vol. 2. Chicago: University of Chicago Press, 1956.

———. *Die Philosophischen Schriften von Leibniz.* Ed. C. I. Gerhardt. 7 vols. Berlin: Weidmann, 1875–1890.

———. Letter to Louis Bourguet. March 22, 1714. In *Die Philosophischen Schriften von Leibniz*, trans. L. Strickland, ed. C. I. Gerhardt, vol. 7. Berlin: Weidmann, 1890.

———. *A Brief Demonstration of a Notable Error of Descartes and Others Concerning a Natural Law.* 1686. In *Philosophical Papers and Letters*, ed. and trans. L. Loemker, vol. 1. Chicago: University of Chicago Press, 1956.

———. "Considerations on Vital Principles and Plastic Natures, by the Author of the System of Pre-Established Harmony." 1705. In *Philosophical Papers and Letters*, ed. and trans. L. Loemker, vol. 2. Chicago: University of Chicago Press, 1956.

———. Fifth Reply to Clarke. Nos. 113–116. 1716. In *Philosophical Papers and Letters*, ed. and trans. L. Loemker, vol. 2. Chicago: University of Chicago Press, 1956.

———. *Philosophical Papers and Letters.* Ed. and trans. L. Loemker. 2 vols. Chicago: University of Chicago Press, 1956.

———. *Principles of Nature and Grace.* 1715. In *Philosophical Papers and Letters*, ed. and trans. L. Loemker, vol. 2. Chicago: University of Chicago Press, 1956.

———. *Reflections on the Common Concept of Justice.* 1702. In *Philosophical Papers and Letters*, ed. and trans. L. Loemker, vol. 2. Chicago: University of Chicago Press, 1956.

———. *Reflections on the Doctrine of a Single Universal Spirit.* 1702. In *Philosophical Papers and Letters*, ed. and trans. L. Loemker, vol. 2. Chicago: University of Chicago Press, 1956.

———. *Specimen Dynamicum.* 1695. In *Philosophical Papers and Letters*, ed. and trans. L. Loemker, vol. 2. Chicago: University of Chicago Press, 1956.

———. "Against Barbaric Physics: Toward a Philosophy of What there actually Is and Against the Revival of the Qualities of the Scholastics and Chimerical Intelligences." In *Philosophical Essays*, ed. and trans. R. Ariew and D. Garber. Indianapolis: Hackett, 1989.

———. *Discourse on Metaphysics.* 1686. In *Philosophical Essays*, ed. and trans. R. Ariew and D. Garber. Indianapolis: Hackett, 1989.

———. *A New System of the Nature and Communication of Substances, and of the Union of the Soul and Body.* 1695. In *Philosophical Essays*, ed. and trans. R. Ariew and D. Garber. Indianapolis: Hackett, 1989.

———. *Philosophical Essays.* Ed. and trans. R. Ariew and D. Garber. Indianapolis: Hackett, 1989.

———. *Principles of Nature and Grace Based on Reason.* In *Philosophical Essays*, ed. and trans. R. Ariew and D. Garber. Indianapolis: Hackett, 1989.

———. *The Principles of Philosophy, or the Monadology.* 1714. In *Philosophical Essays*, ed. and trans. R. Ariew and D. Garber. Indianapolis: Hackett, 1989.

———. *New Essays on Human Understanding.* Ed. and trans. P. Remnant and J. Bennett. Cambridge: Cambridge University Press, 1996.

———. Letter to Louis Bourguet. March 22, 1714. Trans. L. Strickland. Accessed May 6, 2011. http://www.leibniz-translations.com/bourguet1714.htm.

Lennox, J. "The Comparative Study of Animal Development: William Harvey's Aristotelianism." In *The Problem of Animal Generation in Early Modern Philosophy*, ed. J. E. H. Smith, 21–46. Cambridge: Cambridge University Press, 2006.

Lenoir, T. "Kant, Blumenbach, and Vital Materialism in German Biology." *Isis* 71 (1980): 77–108.

———. *The Strategy of Life: Teleology and Mechanics in Nineteenth-Century German Biology.* Cambridge: Cambridge University Press, 1982.

Levere, T. H. *Poetry Realized in Nature: Samuel Taylor Coleridge and Early Nineteenth-Century Science.* Cambridge: Cambridge University Press, 2002.

Linebaugh, P., and M. Rediker, eds. *The Many-Headed Hydra: Sailors, Slaves, Commoners, and the Hidden History of the Revolutionary Atlantic.* Boston: Beacon, 2000.

Lloyd, G. E. R. "The Development of Aristotle's Theory of the Classification of Animals." *Phronesis* 6 (1961): 59–81.

———. "Aristotle's Principle of Individuation." *Mind* 79 (1970): 510–529.

Locke, J. *Of the Conduct of the Human Understanding.* London: W. Bowyer, 1706.

BIBLIOGRAPHY

———. *Johannis Lockii Armigeri Libri IV de Intellectu Humano.* Leipzig: Theophilum Georgi, 1741.
———. Addendum to the *Essay Concerning Human Understanding.* In *Johann Lockens Anleitung des menschlichen Verstandes zur Erkenntnis der Wahrheit,* trans. G. Kypke. Königsberg: J. H. Hartung, 1755.
———. "A Discourse of Miracles." In *Johann Lockens Anleitung des menschlichen Verstandes zur Erkenntnis der Wahrheit,* trans. G. Kypke. Königsberg: J. H. Hartung, 1755.
———. *Elements of Natural Philosophy.* In *The Works of John Locke,* 12th ed., vol. 3. London: C. and J. Rivington, 1824.
———. Locke, J. "Locke's Reply to the Bishop of Worcester's Answer to His Second Letter." In *The Works of John Locke,* 12th ed., vol. 4. London: C. and J. Rivington, 1824.
———. *Some Thoughts Concerning Education.* Ed. R. H. Quick. Cambridge: Cambridge University Press, 1902.
———. *An Early Draft of Locke's Essay, Together with Excerpts from His Journals.* Ed. R. I. Aaron and J. Gibb. Oxford: Clarendon, 1936.
———. Journal entry. November 19, 1677. In *An Early Draft of Locke's Essay, Together with Excerpts from His Journals,* ed. R. I. Aaron and J. Gibb, 99. Oxford: Clarendon, 1936.
———. *The Works of John Locke.* 10 vols. Darmstadt: Scientia, 1963.
———. Letter to Thomas Molyneux. January 20, 1692. In *Dr. Thomas Sydenham (1624–1689): His Life and Original Writings,* by K. Dewhurst, 179–180. Berkeley: University of California Press, 1966.
———. *An Essay Concerning Human Understanding.* 1690. Ed. P. H. Nidditch. Oxford: Clarendon, 1975.
———. *The Correspondence of John Locke.* Ed. E. S. De Beer. 8 vols. Oxford: Clarendon, 1976–1989.
———. *Drafts for the "Essay Concerning Human Understanding" and Other Philosophical Writings.* Vol. 1, *Drafts A and B.* Ed. P. H. Nidditch and G. A. J. Rogers. Oxford: Clarendon, 1990.
Longuenesse, B. *Kant and the Capacity to Judge: Sensibility and Discursivity in the Transcendental Analytic of the "Critique of Pure Reason."* Trans. C. T. Wolfe. Princeton, NJ: Princeton University Press, 1998.
Louden, R. *Kant's Human Being: Essays on His Theory of Human Nature.* Oxford: Oxford University Press, 2011.
Löw, R. *Philosophie des Lebendigen.* Frankfurt: Suhrkamp, 1980.
Magner, L. N. *A History of the Life Sciences.* New York: M. Dekker, 2002.
Magnus, R. *Goethe als Naturforscher.* Leipzig: J. A. Barth, 1906.
Makkreel, R. *Imagination and Interpretation in Kant: The Hermeneutical Import of the "Critique of Judgment."* Chicago: University of Chicago Press, 1990.
———. "Kant on the Scientific Status of Psychology, Anthropology, and History." In *Kant and the Sciences,* ed. E. Watkins, 185–201. Oxford: Oxford University Press, 2001.

Malthus, T. *Essay on the Principle of Population*. In *The Works of Thomas Robert Malthus*, ed. E. Wrigley and D. Souden, vol. 1. London: Pickering, 1798.

Maupertuis, P. L. M. *Discours sur la différente figure des astres avec une exposition des systems de MM. Descartes et Newton*. Paris, 1732; 2nd ed., 1742.

———. "Accord de différentes loix de la nature qui avoient jusque'ice paru incompatibles." In *Mémoires de l'Académie Royal des Sciences*, 417–426. Paris, 1744.

———. "Les loix du mouvement du repos, deduites d'un principe metaphysique, par M. de Maupertuis." 1746. In *Histoire de l'Académie Royale des Sciences et des Belles-Lettres de Berlin*, 267–294. Berlin: Ambroise Haude, 1748.

———. *Essay on Cosmology*. 1750. In *Oeuvres*, vol. 1. Lyon: Frères Bruyset, 1756.

———. *Versuch einer Cosmologie*. Berlin: C. G. Nicolai, 1751.

———. ["Dr. Baumann," pseud.]. *Dissertatio inauguralis metaphysica de universali naturae systemate*. Berlin, 1751.

———. *Oeuvres*. Dresden: C. Walther, 1752.

———. *Essai sur la formation des corps organizes*. 1754. In *Oeuvres*, vol. 2. Lyon: Frères Bruyset, 1756.

———. *Versuch von der Bildung der Körper, aus den Lateinischen des Herrn von Maupertuis übersetzt von einem Freunde der Naturlehre*. Leipzig, 1761.

———. *The Earthly Venus*. Trans. S. Boas. New York: Johnson Reprint, 1966.

———. *Oeuvres*. Ed. G. Tonelli. 4 vols. Hildesheim: G. Olms, 1974.

May, J. A. *Kant's Concept of Geography and Its Relation to Recent Geographical Thought*. Toronto: University of Toronto Press, 1970.

Mayr, E. *This Is Biology: The Science of the Living World*. Cambridge: Cambridge University Press, 1997.

McLaughlin, P. "Kant on Heredity and Adaptation." In *Heredity Produced: At the Crossroads of Biology, Politics, and Culture, 1500–1870*, ed. S. Müller-Wille and H.-J. Rheinberger, 277–291. Cambridge, MA: MIT Press, 2007.

Meier, G. F. *Vernuftlehere*. Halle: J. J. Gebauer, 1752.

———. *Auszug aus der Vernunftlehre*. Halle: J. J. Gebauer, 1760.

Mensch, J. "Kant on Truth." *Idealistic Studies* 34, no. 2 (Fall 2004): 163–172.

———. "Between Sense and Thought: Synthesis in Kant's Deductions." *Epoche: A Journal of the History of Philosophy* 10, no. 1 (Fall 2005): 81–93.

———. "Morality and Politics in Kant's Philosophy of History." In *Toward Greater Human Solidarity: Options for a Plural World*, ed. A. Balslev, 69–85. Kolkata: Dasgupta, 2005.

———. "Kant and the Problem of Idealism: On the Significance of the Göttingen Review." *Southern Journal of Philosophy* 44, no. 2 (Summer 2006): 297–317.

———. "The Key to All Metaphysics: Kant's Letter to Herz, 1772." *Kantian Review* 12, no. 2 (2007): 109–127.

———. "Material Unity and Natural Organism in Locke." *Idealistic Studies* 40, no. 1–2 (Fall 2010): 149–164.

———. "Intuition and Nature in Kant and Goethe." *European Journal of Philosophy* 19, no. 3 (Fall 2011): 431–453.

———. "Understanding Affinity: Locke on Generation and the Task of Classification." *Locke Studies* 11 (2011): 49–71.

Menzer, P. *Kants Lehre von der Entwicklung in Natur und Geschichte*. Berlin: G. Riemer, 1911.

Meynell, G. G. "A Database for John Locke's Medical Notebooks." *Medical History* 42 (1997): 473–486.

———. "Locke as a Pupil of Peter Stahl." *Locke Studies* 1 (2001): 221–227.

———. "John Locke and the Preface to Thomas Sydenham's *Observationes Medicae*." *Medical History* 50 (2006): 93–110.

Miller, E. *The Vegetative Soul: From Philosophy of Nature to Subjectivity in the Feminine*. Albany: SUNY Press, 2002.

Milton, J. R. "Locke, Medicine, and the Mechanical Philosophy." *British Journal for the History of Philosophy* 9 (2001): 221–243.

Morton, A. G. *History of Botanical Science: An Account of the Development of Botany from Ancient Times to the Present Day*. London: Academic, 1981.

Müller-Sievers, H. *Self-Generation: Biology, Philosophy, and Literature around 1800*. Stanford, CA: Stanford University Press, 1997.

Needham, J. T. "A Summary of some late Observations upon the Generation, Composition, and Decomposition of Animal and Vegetable Substances." *Philosophical Transactions of the Royal Society* 45 (1748): 615–666.

Newman, W. *Atoms and Alchemy: Chymistry and the Experimental Origins of the Scientific Revolution*. Chicago: University of Chicago Press, 2006.

Newton, I. *Mathematical Principles of Natural Philosophy*. Trans. A. Mott. Berkeley: University of California Press, 1934.

Nieman, S. *The Unity of Reason: Rereading Kant*. New York: Oxford University Press, 1994.

Nova Acta Eruditorum. Unsigned review of *Gothofredi Guillelmi Leibnitii, Opera Omnia* (1768), by G. W. Leibniz, ed. L. Dutens. October 1768, 433–449.

———. Unsigned review of *Historia critica philosophiae*, 2nd ed., by J. J. Brucker. April 1769, 156–173.

Park, P. K. J. *Africa, Asia, and the History of Philosophy: Racism in the Formation of the Modern Canon, 1780–1830*. Albany: SUNY Press, 2013.

Piché, C. "The Precritical Use of the Metaphor of Epigenesis." In *New Essays on the Precritical Kant*, ed. T. Rockmore, 182–200. New York: Humanity Books, 2001.

Preus, A. *Science and Philosophy in Aristotle's Biological Works*. Hildesheim: G. Olms, 1975.

Quarfood, M. *Transcendental Idealism and the Organism: Essays on Kant*. Stockholm: Almqvist and Wiksell International, 2004.

Ramsay, T. "To the Lovers of Natural History." *Scots Magazine* 34 (1774): 174–175.

Réaumur, R. "De la formation et de l'acroissement des coquilles des animaux tant terrestes qu'aquatiques, soit de mer soit de rivière." In *Mémoires de l'Académie Royale des Sciences*, 364–400. N.p., 1709.
———. *Mémoires pour server à l'histoire des insectes*. Paris: De l'imprimerie royale, 1734.
———. "Quatriéme mémoire: Esquisse des amusemens philosophiques que les oiseaux d'une basse-cour ont à offrir." In *Art de faire éclorre et d'élever en toute saison des oiseaux domestiques de toutes especes*, vol. 2. Paris: Imprimerie Royale, 1749.
———. *The Art of Hatching and Bringing Up Domestic Fowls by Means of Artificial Heat*. Trans. A. Trembley. London, 1750.
Reill, P. "Between Mechanism and Romantic *Naturphilosophie*: Vitalizing Nature and Naturalizing Historical Discourse in the Late Enlightenment." In *Regimes of Description: In the Archive of the Eighteenth Century*, ed. J. Bender and M. Marrinan, 153–174. Stanford, CA: Stanford University Press, 2005.
Richards, R. J. "Kant and Blumenbach on the *Bildungstrieb*: A Historical Misunderstanding." *Studies in the History and Philosophy of Biological and Biomedical Science* 31 (2000): 11–32.
———. *The Romantic Conception of Life: Science and Philosophy in the Age of Goethe*. Chicago: University of Chicago Press, 2002.
Ripstein, A. *Force and Freedom: Kant's Legal and Political Philosophy*. Cambridge, MA: Harvard University Press, 2009.
Robert, J. *Embryology, Epigenesis, and Evolution: Taking Development Seriously*. Cambridge: Cambridge University Press, 2004.
Roe, S. *Matter, Life, and Generation: Eighteenth-Century Embryology and the Haller-Wolff Debate*. Cambridge: Cambridge University Press, 1981.
———, ed. and trans. *The Natural Philosophy of Albrecht von Haller*. New York: Arno, 1981.
Roger, J. *Buffon: A Life in Natural History*. Trans. S. Bonnefoi. Ithaca, NY: Cornell University Press, 1997.
———. *The Life Sciences in Eighteenth-Century French Thought*. Trans. R. Ellrich. Stanford, CA: Stanford University Press, 1997.
Rogers, G. A. J. *Locke's Enlightenment: Aspects of the Origin, Nature and Impact of His Philosophy*. Hildesheim: G. Olms, 1998.
Ruestow, E. G. "Images and Ideas: Leeuwenhoek's Perception of the Spermatozoa." *Journal of the History of Biology* 16 (1983): 185–224.
Rutherford, D. *Leibniz and the Rational Order of Nature*. Cambridge: Cambridge University Press, 1995.
Sachs, J. von. *History of Botany (1530–1860)*. Trans. H. E. F. Garnsey. Oxford: Clarendon, 1906.
Sallis, J. *The Gathering of Reason*. 2nd ed. Albany: SUNY Press, 2005.
Schaffer, S. "Godly Men and the Mechanical Philosophers: Souls and Spirits in Restoration Natural Philosophy." *Science in Context* 1 (1987): 55–85.

Schelling, F. W. J. *The Philosophy of Art.* 1801. Trans. D. Stott. Minneapolis: University of Minnesota Press, 1989.

———. *First Outline of a System of the Philosophy of Nature.* 1799. Trans. and ed. K. R. Peterson. Albany: SUNY Press, 2004.

Schiller, F. *On the Aesthetic Education of Man.* Trans. R. Snelling. Mineola, NY: Dover, 2004.

Schlegel, F. *On the Study of Greek Poetry.* Trans. S. Barnett. Albany: SUNY Press, 2001.

Schofield, R. E., and D. G. C. Allan. *Stephen Hales, Scientist and Philanthropist.* London: Scolar, 1980.

Scholz, H. *Die Hauptschriften zum Pantheismusstreit.* Berlin: Reuther and Reichard, 1916.

Schönfeld, M. *The Philosophy of the Young Kant: The Precritical Project.* Oxford: Oxford University Press, 2000.

Schrecker, P. "Leibniz and the Timaeus." *Review of Metaphysics* 4 (1951): 495–505.

Schultz, J. "*Institutiones Logicae et Metaphysica* by Jo. Aug. Hen. Ulrich." In *Kant's Early Critics: The Empiricist Critique of the Theoretical Philosophy,* trans. B. Sassen. Cambridge: Cambridge University Press, 2000.

Seamon, D., and A. Zajonc, eds. *Goethe's Way of Science: A Phenomenology of Nature.* Albany: SUNY Press, 1998.

Shaw, B. W. "Function and Epigenesis in Kant's *Critique of Pure Reason.*" Master's thesis, University of Georgia, 2003.

Shell, S. M. *The Embodiment of Reason: Kant on Spirit, Generation, and Community.* Chicago: University of Chicago Press, 1996.

———. "Kant as Propagator: Reflections on *Observations on the Feeling of the Beautiful and Sublime.*" *Eighteenth-Century Studies* 35, no. 3 (2002): 455-468.

———. *Kant and the Limits of Autonomy.* Cambridge, MA: Harvard University Press, 2009.

Sloan, P. R. "John Locke, John Ray, and the Problem of the Natural System." *Journal of the History of Biology* 5 (1972): 1–53.

———. "The Idea of Racial Degeneracy in Buffon's *Histoire naturelle.*" *Studies in Eighteenth-Century Culture* 3 (1973): 293–321.

———. "The Buffon-Linnaeus Controversy." *Isis* 67 (1976): 356–375.

———. "Preforming the Categories: Eighteenth-Century Generation Theory and the Biological Roots of Kant's A Priori." *Journal of the History of Philosophy* 40 (2002): 229–253.

Smith, J. E. H. *Divine Machines: Leibniz and the Sciences of Life.* Princeton, NJ: Princeton University Press, 2011.

Smith, N. K. *Commentary to Kant's "Critique of Pure Reason."* 1918, rev. ed., 1923. Reprint, Atlantic Highlands, NJ: Humanities Press, 1984.

Sommer, A., and W. Conze. "Rasse." In *Geschichtliche Grundbegriffe: Historisches*

Lexicon zur politisch-sozialen Sprache in Deutschland, ed. O. Brunner, W. Conze, and R. Koselleck, 5:137–141. Stuttgart: Klett-Cotta, 1984.
Sturm, T. "Kant on Empirical Psychology." In *Kant and the Sciences*, ed. E. Watkins, 163–184. Oxford: Oxford University Press, 2001.
Talbott, S. "Getting Over the Code Delusion." *New Atlantis* 28 (2010): 3–27.
Terrall, M. *The Man Who Flattened the Earth: Maupertuis and the Sciences in the Enlightenment*. Chicago: University of Chicago Press, 2002.
Tetens, J. N. *Über die allgemeine spekulativische Philosophie*. Bützow: Berger und Boednerschen Buchhandlung, 1775.
———. *Philosophische Versuche über die menschliche Natur und ihre Entwickelung*. Leipzig: Weidmanns Erben und Reich, 1777.
———. *Einleitung zur Berechnung der Leibrenten und Anwartschaften*. 2 vols. Leipzig: Weidmanns Erben und Reich, 1785–1786.
Tonelli, G. Introduction to *Oeuvres*, by P. L. M. Maupertuis, 1:xi–lxxxiii. Hildesheim: G. Olms, 1974.
———. "Leibniz on Innate Ideas and the Early Reactions to the Publication of the *Nouveaux essais* (1765)." *Journal of the History of Philosophy* 12 (1974): 437–454.
Trembley, A. *Mémoires pour servir a l'histoire d'un genre de polypes d'eau douce*. Paris, 1744.
Uebele, W. *Johann Nicolaus Tetens nach seiner Gesamtentwicklung betrachtet, mit besonderer Berücksichtingung des Verhältnisses zu Kant*. Berlin: Reuther and Reichard, 1911.
Walmsley, J. "Morbus—Locke's Early Essay on Disease." *Early Science and Medicine* 5 (2000): 366–393.
———. "Morbus, Locke and Boyle: A Response to Peter Anstey." *Early Science and Medicine* 7 (2002): 378–397.
Watkins, E. "The 'Critical Turn': Kant and Herz from 1770 to 1772." *Proceedings of the Ninth International Kant Congress* 2 (2001): 69–77.
———. *Kant's "Critique of Pure Reason": Background Source Materials*. Cambridge: Cambridge University Press, 2009.
Warda, A. *Immanuel Kants Bücher*. Berlin: M. Breslauer, 1922.
Westfall, R. S. "Biology and the Mechanical Philosophy." In *The Construction of Modern Science: Mechanisms and Mechanics*, 82–104. Cambridge: Cambridge University Press, 1977.
Wilkie, J. S. "Preformation and Epigenesis: A New Historical Treatment." *History of Science* 6 (1967): 138–150.
Winter, A. "Selbst Denken-Antinomien-Schranken Zum Einfluß des späten Locke auf die Philosophie Kants." *Aufklärung* 1 (1986): 27–66.
Wolff, C. *Elementa matheseos universae*. Vol. 1. Halle, 1713–1715.
———. *Allerhand nützliche Versuche*. Halle, 1721.
———. *Vernünftige Gedanken von der Wirkungen der Natur*. Magdeburg, 1723.
———. Preface to *Statick der Gewächse oder angestelte Versuche mit dem Saft*

in *Pflanzen und ihren Wachstum*, by S. Hales. Halle: Rengerischen Buchhandlung, 1748.

Wolff, C. F. "Von der eigenthümlichen und wesentlichen Kraft der vegetabilischen sowohl als auch der animalischen Substanz." In *Zwo Abhandlungen über die Nutritionskraft welche von der Kayserlichen Akademie der Wissenschaft in St. Petersburg den Preis getheilt haben*. St. Petersburg: Kayserliche Akademie der Wissenschaften, 1789.

———. *Theoria generationis*. 1759. Facsimile reprint. Hildesheim: G. Olms, 1966.

———. *Theorie von der Generation in zwo Abhandlungen erklärt und beweisen*. 1764. Facsimile reprint. Hildesheim: G. Olms, 1966.

Wubnig, J. "The Epigenesis of Pure Reason: A Note on the *Critique of Pure Reason*, B, sec. 27, 165-168." *Kant-Studien* 60, no. 2 (1969): 147-152.

Zammito, J. H. *The Genesis of Kant's "Critique of Judgment."* Chicago: University of Chicago Press, 1992.

———. *Kant, Herder, and the Birth of Anthropology*. Chicago: University of Chicago Press, 2002.

———. "'This Inscrutable *Principle* of an Original *Organization*': Epigenesis and 'Looseness of Fit' in Kant's Philosophy of Science." *Studies in History and Philosophy of Science* 34 (2003): 73-109.

———. "Kant's Early Views on Epigenesis. The Role of Maupertuis." In *The Problem of Animal Generation in Early Modern Philosophy*, ed. J. E. H. Smith, 317-354. Cambridge: Cambridge University Press, 2006.

Zöller, G. "Kant on the Generation of Metaphysical Knowledge." In *Kant: Analysen-Probleme-Kritik*, ed. H. Oberer and G. Seel, 71-90. Wurzburg: Königshausen and Neumann, 1988.

———. "Kant's Political Anthropology." *Kant Jahrbuch* 3 (2011): 131-161.

———. "Ursprung: Kants kritische Originalität." In *Urworte: Zur Geschichte und Funktion erstbegründender Begriffe*, ed. M. Ott and T. Döring, 121-134. Munich: W. Funk, 2012.

———. "Between Rousseau and Freud: Kant on Cultural Uneasiness." In *New Studies on Kant*, ed. P. Muchnik. Cambridge: Cambridge Scholars, 2013.

Zuckert, R. *Kant on Beauty and Biology: An Interpretation of the "Critique of Judgment."* Cambridge: Cambridge University Press, 2007.

Zweig, A., trans. and ed. *Correspondence*. The Cambridge Edition of the Works of Immanuel Kant. Cambridge: Cambridge University Press, 1999.

Index

Abrams, Meyer, 219
adaptation of species, 11, 13, 14, 102–4, 106
adapted traits, 102–3, 150
Adickes, Erich, 83, 193, 195
albino, 97, 179
Allan, D. G. C., 173, 175, 219
Allison, Henry, 203–4, 213, 219
Ameriks, Karl, 206, 219
analysis situs, 46, 66, 182, 185, 189–90, 224
Anaxagoras, 177–78. *See also* homoeomeries
animalcules, 31–33, 37, 156, 169, 171, 178. *See also* Leeuwenhoek, Anton
animalculism, 31, 37, 169, 171, 187, 199. *See also* spermist theory of generation
Anlage, 7, 103, 106, 112, 136, 202–3. *See also* disposition
Anstey, Peter, 163, 165–69, 219–20
anthropology, 9, 66, 96, 105, 115, 120–21, 186, 189, 192, 200, 203, 206, 215
aphids, 32, 170, 172. *See also* parthenogenesis; virgin birth
apperception, 12–14, 32, 94, 108, 118–19, 130, 133–34, 137, 158, 207–8
aptitude, 10, 32, 81, 94, 107–8, 153, 160, 213
archaeology, 14, 150, 161
archetypal, 15, 26, 88, 150–52, 168, 192, 206

architect, 19, 127, 140
architectonic, 125, 127, 139. *See also Bauplan*
Aristotle, 5, 7, 17–18, 30, 33, 65–66, 74, 82, 83, 89–90, 109, 122, 126, 146, 157, 162–65, 172–73, 177, 194, 209, 220, 223
astronomers, 85, 194
atomism, 164, 169, 223
Aucante, Vincent, 165, 220
ausarten, -ung, 101, 150. *See also* degeneration
auswickeln, -ung, 6, 62–63, 102–3, 150, 187–88, 200–201. *See also* unfolded

Basedow, Johann, 95, 105–6, 203. *See also* Philanthropinum School
Baumann thesis, 180–81, 186. *See also* Maupertuis, Pierre-Louis
Baumgarten, Alexander, 52, 82–83, 114, 193–94, 206–7, 220
Bauplan, 125, 131, 133, 139–40. *See also* architectonic
Beck, Lewis White, 187, 193–94, 204–5, 207, 213, 219–20
Beguelin, Nicholas, 193, 220
Beiser, Frederick, 220
Bernasconi, Robert, 199–200, 202, 220
Bernier, François, 198, 220
Bildung, aus-, 96, 105–6, 112, 116, 180–81, 186, 210. *See also* cultivation

INDEX

Bildungstrieb, 36, 144, 155, 160–62, 210, 216, 220. *See also* formative force
biological: affinity, 15, 152; analogs, 81, 124; epigenesis, 7, 153, 214; generation, 20, 33, 53, 99, 159, 179, 194; grounds, 15, 152; models, 111, 152, 159; organisms, 3, 159; origin, 7, 52–53, 58, 60–61, 72, 99; vocabulary, 89, 159
Birken-Bertsch, Hanno, 190, 220
birth: certificate of 131–32, 136; of concepts, 136; of metaphysics, 129; of morality, 213; self-, 13, 133, 158, 214; of soul, 82, of species, 31, 40, 50; virgin, 32, 81, 213
birthplace: of cognition, 131; of concepts, 79, 135–37; of faculties, 135; of moral law, 213; of rules, 13
blackbirds, 97, 198
Blair, Patrick, 37, 43, 174, 220
blending: and joint inheritance, 7, 82, 96; and progeny, 7, 180, 199; racial, 99, 101, 199
Blumenbach, Johann, 36, 144, 153, 160–62, 210, 216, 220
Boerhaave, Hermann, 38, 53, 65, 69, 157
Bonnet, Charles 11, 63, 68, 111–13, 119, 142, 156, 160, 171–72, 201–2, 205, 221, 224
Borelli, Giovanni, 68, 163
botanical models, 21, 105, 131
botany, 16, 20–21, 23, 100, 163, 165, 167, 174–75, 183, 185, 220
Bourguet, Louis, 31, 35, 170–71, 173, 175–77, 221, 224. *See also* crystal growth; intussusception
Bowler, Peter, 156–57, 221
Brandt, Reinhardt, 191, 221
breeders, 17, 97, 198
Brucker, Johann, 192
Buffon, George Leclerc, 4–7, 10, 16–17, 35–36, 38–50, 53–54, 57, 59–60, 62–63, 66–68, 80, 82, 96, 98–102, 104, 119, 153, 156–57, 160, 173, 175–83, 185–90, 197–202, 221–23, 226
Buroker, Jill Vance, 190, 223

Calvin, Jean, 17, 29, 162, 223
Camerarius, Rudolf, 31, 38, 175, 223
Carl, Wolfgang, 196, 204, 206, 223
Cartesian, 3, 33, 40, 58
Castillon, Jean, 193, 223

Cesalpino, Andreas, 163, 223
chemistry, 20, 35, 44–45, 82, 163–64, 221, 223
Clericuzio, Anthony, 164, 223
climate, 4, 11, 14, 48, 96–99, 101–4, 198, 200, 202
concepts: empirical, 71, 74–75, 78–81, 85–86, 113, 122, 133–35; epigenesis of, 10, 13, 122, 124, 193, 212, 214; innate, 33, 89, 193, 209; intellectual, 8, 10, 12–13, 74, 78–80, 83, 86–87, 90–92, 108–9, 112–13, 117–18, 122, 124, 127, 131–32, 135–37, 195–96, 204; origin of, 122–23, 136–37, 193; originally acquired, 78–81, 84–85, 107, 122, 124; pure, 75, 78, 85, 88–89, 117, 133–36; of reason, 104, 119–20, 127, 136, sensitive, 79, 83, 190; subreptive, 8, 70–72, 189–90; surreptitious, 71–72, 84, 95, 190
concreationism, 82, 194, 201
Condillac, Etienne 113, 201
convention of matter, 19, 47, 164
Cooper, John, 162
corpuscles, 18, 164, 167–68, 223
corpuscular hypothesis, 32, 18–21, 28–29, 163–66, 168, 221
cosmological theories, 6–7, 41, 52–56, 58–60, 63, 72, 186, 200
cosmology, 53, 56, 60, 114, 184–86
crows, 97, 198
Crusius, Christian, 52, 83, 87, 89, 159, 195
crystal growth, 35, 42, 44, 153, 173, 177–78. *See also* Bourguet, Louis; intussusception
cultivation, 105–7, 116, 202–3, 210. See also *Bildung, aus-*

Darwin, Charles, 9, 15, 83, 151–52, 158, 160, 162, 194, 198, 217, 223, 225
Deason, Gary, 162, 223
deduction of the categories of experience, 2, 13, 93, 114, 117, 119, 123–25, 131–35, 137–38, 159, 196, 207–8, 223
degeneration, 4, 48–50, 98, 101, 150, 183, 198, 200, 222
De Risi, Vincenzo, 182, 224
Descartes, René, 17, 19–20, 22, 28, 33, 53, 80, 163, 165, 172, 183, 193, 220, 223–24

240

descent, 15, 30, 79, 127, 132, 135–36, 152, 167
Des Chene, Dennis, 164, 224
Detlefsen, Karen, 157, 224
développer, 46, 178, 187, 200–202. See also unfolded
De Vleeschauwer, Hermann, 204, 224
dialectic, 5, 133, 147, 149, 190, 220
disposition, 10–12 32, 81, 94, 96, 98–99, 101–4, 106–8, 112, 136, 138, 200–204, 213. See also *Anlage*; germs; predispositions
dissection of the understanding, 113, 135
distribution of species, 39, 67, 97–99, 102, 104, 106, 183
divergence, 107, 130, 187, 192
dogmatic metaphysics, 99, 103, 126, 128–29, 133, 195, 217
Dörflinger, Bernd, 224
Downing, Lisa, 168, 224
Duchesneau, François, 159–61, 173, 224
Dutens, Louis, 171, 182, 189, 192

Eberhard, Johann, 108, 192, 204, 219
eclectic, 51, 53, 110
eclecticism, 52, 60, 166
education, 95, 155, 169, 199, 203
educt, 109, 194
emboîtement, 156, 161, 187. See also encasement theory of generation; Russian doll model of generation; unfolded
embryo, 4–5, 23, 37, 43, 46, 99, 112, 156, 166–67, 180, 182, 188, 200
embryogenesis, 1, 3–5, 7, 15, 53, 96–99, 102
embryological, 1, 45, 53, 70, 156–58, 160, 162, 172, 223
emergence, 112, 150, 153, 161, 224
emergent properties, 12, 15, 36, 108, 154, 158, 160–61, 215–17
empiricism, 73, 118–19, 159, 206
empiricist, 32, 88, 123, 135, 205, 208
empiricists, 10, 73–74, 83, 114, 118–20, 126, 132, 138, 209
encasement theory of generation, 3, 23, 31, 37, 62, 81, 112, 150, 156, 161, 187–88. See also Russian doll model of generation
Engel, Johann, 197, 224
entelechy, 17, 29–30, 33–34, 43, 45, 178
Entstehung, 112, 191, 221

entwickeln, 9, 102, 136, 187, 201, 213
Entwicklung, -elung, 106, 112, 188, 200–202, 204, 215, 225
Epicurus, 126, 169
epigenesis: Aristotle on, 5, 157; biological, 7, 82, 144, 153, 157–59, 216, 230; Buffon on, 5, 7, 43, 82, 156–57; of concepts, 10, 89, 108, 204, 214; Harvey on, 5, 7, 11, 166; mechanical, 5, 43, 156–57; as model for cognition, 9, 80–83, 109, 153, 158, 209, 214–15; versus preexistence theory, 5, 82–83, 99, 156–57, 188, 194; of reason, 8–9, 13, 15, 81, 89, 124, 138–39, 141, 144, 158–59, 193, 212, 214–15; C.F. Wolff on, 5, 158, 201
epigenetics, 15, 154, 217
erzeugen, -ung, 6, 62–63, 100, 102, 103, 122, 129. See also generation: organic
Evo-Devo, 182, 223
evolution, 13, 110–12, 155, 158, 160–61, 178, 183, 187, 201–2, 210

faculties of cognition, 9, 13, 32, 75, 77, 81, 92, 94, 108, 113, 115–18, 127, 129, 131, 133–35, 139, 158, 161, 186, 206, 210, 213, 215
fallacy of subreption, 87, 95. See also concepts
Farley, John, 224
fate of reason, 71, 73, 128, 130, 140, 149, 219–20
Fichte, Johann, 220
Fischer, Karl, 197, 224
Fisher, Mark, 188, 225
foetus, 37, 43–44, 167, 181, 189, 222
formative force, 9, 17–19, 116–18, 150, 153, 155, 160–61, 163–65, 207, 224
Förster, Eckhart, 196, 207, 223, 225
freedom, 2, 107–9, 111–12, 139, 153, 188, 203–4, 214–15
Friedman, Michael, 186, 225

Gadamer, Hans Georg, 210, 225
Galileo, 17
Garber, Daniel, 170, 172, 225
Garve, Christian, 111
Gasking, Elizabeth, 160, 169, 177, 225
genealogical, 12, 14, 48–49, 101, 121, 124, 131, 133, 135–36, 141, 149, 215
genealogy, 5, 12–13, 47, 50, 114, 121, 123, 129, 185

INDEX

genera, 27, 50, 149–50, 163
generatio aequivoca, 127, 138
generation: epigenetic, 7, 123–24, 129, 153, 158, 212; fermentation model of, 3, 20, 22, 28, 63, 161, 165; of knowledge, 8–9, 80, 83, 109, 124; mechanical models of, 7, 28, 34, 63, 97, 156; of mind, 81, 83, 158, 214; of moral exemplars, 79; organic, 5–6, 9, 16–17, 20, 23, 27–28, 33, 35, 59–64, 72, 81, 99, 103–5, 107–8, 112, 129, 141, 214; of organisms 2–4, 16, 21–22, 37–38, 40–41, 44, 53, 82–83, 110–11, 143–44, 159, 162, 165, 174, 194; preexistence theory of, 28–29, 31–33, 37, 42–43, 62, 82, 99, 161, 180, 187, 201; of racial traits, 11, 96–112, 200; spontaneous, 7, 82, 162, 212; supernatural, 6, 62–63, 81
generic preformation of species lines, 6–8, 10, 46, 62, 144, 157, 214
genes, 15, 158, 217
genotype, 158
Genova, A. C., 159, 225
geographic distribution, 11, 48, 97–99, 101–2, 104, 198, 212
geography, 5, 9–11, 14, 59–60, 66, 95–96, 182, 186, 197–98, 202, 217, 220
geology, 6, 59, 185, 216
geometry, 39, 46, 60, 66–67, 75–79, 136, 182, 185, 189–90, 195, 224
germs: as adaptive response, 103, 107, 136, 200; and categories of experience, 136; for cultivation, 105, 202; epigenetically produced, 12–13, 136, 204; preformed, 10–11, 63, 97, 104, 156, 160, 179, 200–201, 204; of reason, 105, 202; teleological approach toward, 14, 104. *See also* disposition
Geschlecht, 175, 189, 223
geschwängert, 13, 133
Gigante, Denise, 210, 225
Goethe, Johann, 14, 151–52, 161–62, 178, 193, 211–12, 216–17, 221, 225–26
Gough, John, 165, 167, 226
Grier, Michelle, 190, 226
Guerlac, Henri, 173–74, 226
Guyer, Paul, 194, 212, 226

Haffner, Thomas, 158, 214, 226
Hall, Thomas, 177, 226

Haller, Albrecht von, 35, 43, 59–60, 72, 157, 172–73, 177–78, 186, 188, 224, 226. *See also* irritability
Hamann, Georg, 112, 206
Hankins, Thomas, 172–73, 226
Harvey, William, 5, 7, 11, 20, 22, 157–58, 166, 169, 226
Herder, Johann, 13, 51–52, 112, 155, 177, 183, 188–89, 191–92, 194, 197, 200, 206, 210, 227
Hegel, Georg, 206, 214
Helmont, Jan van, 165–66, 171
Henrich, Dieter, 207–8, 227
Henry, John, 162, 164, 227
Hercules, 147–48
Herder, Johann, 13, 51–52, 112, 155, 183, 188–89, 191–92, 194, 197, 200, 206, 210, 227
heredity, 82, 101, 160, 199, 225
heritable traits, 14, 97–98, 180
hermaphroditic, 37
Herz, Marcus, 83, 88–92, 111–12, 138, 195–96, 227
Hesiod, 213
heuristic guides for thought, 2, 8, 14–15, 151–52, 161, 199
homoeomeries, 177. *See also* Anaxagoras
horticulturalists, 17, 23
Hume, David, 13, 52, 73, 87–88, 126, 133–35, 138, 159, 161,176, 191, 193–94, 212–13, 219–20
Huneman, Phillipe, 188, 227
Hutcheson, Francis, 73
Huygens, Christian, 182, 189
hybrids, 4, 23, 25, 37, 48–49, 164
hydra, 3–4, 11, 43, 178, 201
hylomorphism, 162
hylozoism, 65, 181

illusions of reason, 12, 95, 120, 147, 149, 190, 194
imagination, 38, 50, 57, 110, 114, 116–19, 122, 134–35, 160–61, 207
immaculate conception, 199
impregnation, 13, 37, 133
incongruent counterparts, 67, 69, 189–90
individuation, 163, 170
induction, 10, 146, 193, 209
inductive method, 39, 146–47
influx physico theory, 83, 89

INDEX

infusoria, 30, 82
Ingensiep, Hans, 158–59, 214
inheritance: and birthright, 13, 131; of traits, 7, 11, 14, 17, 43–44, 96–98, 102, 106, 178
innate: character of good, 204; concepts, 9, 33, 108, 193, 209; germs, 105; ideas, 10, 13, 32, 73, 78, 80, 192; knowledge, 170; laws, 81, 107; mental predisposition, 107–8; moral feeling, 203; orientation, 107; truths, 108
inorganic, 19, 35, 44, 102, 161, 216
intellect, 75–76, 78–80, 84–89, 116–17, 126, 190, 195–96, 214
interfertility criterion, 10, 49, 100, 104, 161
intussusception, 35, 128, 173, 209. *See also* Bourguet, Louis; crystal growth
irritability, 35, 72, 177. *See also* Haller, Albrecht von

Jacobi, Heinrich, 175, 212, 223
Jacquette, Dale, 206, 227
judgments: aesthetic, 121; empirical, 80, 84, 133; logical, 10, 12, 77, 90–91, 113, 117–18, 133, 136–37, 141, 161, 191; reflective, 2, 14, 142–43, 152, 207, 216; table of, 136–37, 141, 215

Kant, Immanuel, works of: *Critique of Judgment*, 2, 142, 144, 150–51, 156, 158–59, 206, 211, 215; *Critique of Practical Reason*, 107, 139, 212; *Critique of Pure Reason*, 9, 12–13, 51, 53, 90–92, 95, 108, 110, 113–14, 118, 120–25, 128, 129, 131, 135, 140, 142, 146, 149, 156, 158–59, 190, 192–95, 203–4, 207, 208; *Dreams of a Spirit-Seer*, 57, 64, 66, 69–70, 73–74, 86, 157, 184, 191; *Groundwork of the Metaphysics of Morals*, 213; *Inaugural Dissertation*, 52, 60, 75–76, 78, 80, 83–88, 90, 93–94, 107, 111, 116, 132, 179, 186, 190, 195, 205; *Of the Different Races* (1775/77), 95–104, 199–203; *Only Possible Argument*, 6–7, 14, 61, 65, 81–82, 87, 102–3, 142, 144, 200
Kästner, Abraham, 59, 178, 187–89, 197–200, 221
Keim, *-en*, 9, 103, 105–6, 129, 136, 155, 214. *See also* germs
Kepler, Johannes, 52

Kitcher, Patricia, 204, 229
Kleingeld, Pauline, 202, 229
Kraft, 9, 116, 118, 207, 217. *See also* formative force
Krug, Wilhelm, 210, 229
Kruk, Remke, 165, 229
Kuehn, Manfred, 52, 183–85, 187, 191, 229
Kusch, Martin, 206, 230
Kypke, Georg, 73, 191

Lagier, Ralph, 199–200, 230
Lamark, Jean-Baptiste, 198
Lambert, Johann, 75–76, 83–88, 110, 112, 194, 205, 220, 230
Larrimore, Mark, 199, 230
Lavater, Johann, 201, 205, 221
Lavoisier, Antoine-Laurent, 174, 216, 226
Laywine, Alison, 186, 194, 230
Leeuwenhoek, Anton, 3, 22, 29–32, 37, 39, 54, 156, 166, 169–71, 174, 178
Leibniz, Gottfried, 8, 13, 28–35, 41–43, 45–46, 52, 54, 66–67, 69, 73–74, 76, 80–81, 93, 108–10, 122–23, 126, 138, 142, 159–61, 169–72, 175, 177, 181–82, 186–87, 189–93, 204–5, 209, 213, 220, 224–25, 230–31
Lennox, James, 157, 231
Lenoir, Timothy, 161, 215–16, 231
Linnaeus, Carl, 3–4, 16, 47, 100–101, 178, 182–83
Locke, John, 8, 16–18, 20–29, 33, 47, 53, 73–76, 78, 80–81, 83, 89, 95, 109–10, 113, 118–19, 121–23, 126, 132, 159, 163, 165–70, 172, 191–93, 197, 205, 209, 220–21, 223–24, 226, 231–32

Makkreel, Rudolf, 206, 232
Malebranche, Nicholas, 62, 83, 89, 156, 187
Malpighi, Marcello, 29–31, 169
Malthus, Thomas, 158, 233
materialism, 33, 65, 120, 161, 168, 201
mathematician, 75, 181
mathematics, 26, 39–40, 46, 54, 57–58, 88, 176, 195, 209
Maupertuis, Pierre-Louis, 4, 43–45, 52–56, 58–60, 62–63, 65, 68, 82, 96–99, 101–3, 153, 160, 178–81, 183–84, 186, 198, 200, 219
Mayr, Ernst, 158, 233
McLaughlin, Peter, 199, 233

INDEX

mechanical: approach to nature, 1, 6, 8, 11, 17, 19–20, 23, 28, 34–36, 56–57, 62, 68, 72, 104, 143, 149, 164–65, 170, 173; epigenesis, 5, 42–43; philosophy, 3, 17, 20, 33; theory of generation, 5–8, 19, 21–23, 28, 34, 42–43, 45, 62–65, 72, 81–82, 144, 156, 161

mechanism, 2–3, 13–14, 17, 34–35, 63, 65, 69, 97, 107, 111, 163, 176

Meier, Georg, 207, 233

Mendelssohn, Moses, 85, 111, 191

Menzer, Paul, 188, 200, 215, 234

metamorphosis, 15, 21, 151–52, 166, 211, 226

metaphysical: account of cognition, 7, 12, 108, 124, 144, 153, 158–59, 214–15; knots, 64, 66, 188; mechanics, 55; unity of organism, 30–31, 33–34, 162, 170

metaphysicians, 54, 66, 71, 85, 111, 115, 124, 159, 205, 215

metempsychosis, 31, 171

Meynell, G. G., 165–66, 234

microscopists, 29–32, 169–70

Milton, John, 165–66, 234

molds for species lines, 5, 42, 45–46, 48, 59, 67–68, 98, 153, 156, 167, 176–78, 198

molecules, 41–43, 45–46, 98, 153, 177–78, 187

Molyneux, William, 22, 26, 166, 168, 191

monads, 29–33, 42, 45, 60, 68, 71, 169–70, 177, 181, 187

monogenesis of species, 11, 14, 161

Montesquieu, Charles-Louis de, 73

morality, 26, 73, 142, 210, 213–14

morphogenesis, 46, 160

morphology, 162, 217

Morton, A.G., 163, 175, 183, 234

Moscati, Pietro, 9, 95, 104–5

Müller-Sievers, Helmut, 159, 234

Naturanlage, 129–30, 202. *See also* disposition

Needham, John, 179

Newman, William, 164, 168, 224

nominalism, 27, 58, 168–69

ontology, 18–20, 28, 85, 114, 124, 164–65, 168–69, 224

origin: biological, 7, 52–53, 58, 60–61, 72, 99, 194; of concepts, 75, 78–79, 88–90, 107, 122–24, 126, 131–36, 196, 207; cosmological, 32, 52–54, 56, 58–60, 72; epigenetic 90, 107–9, 113–14, 116, 130, 133–34, 207; geographical, 49–50, 96, 98, 106, 198; of ideas, 8, 32, 53, 73–74, 76, 79–80, 83, 109, 115–16, 126, 159, 192; of knowledge, 32, 53, 72, 74, 79–80, 109, 112, 119, 121, 124, 126, 139, 208; of metaphysics, 73, 127; questions of 2, 6, 47, 52, 75, 81–82, 123–24, 132–33, 136, 195; of races, 51, 97–99; of reason, 12, 133, 203; of species, 28, 48, 50, 96, 98, 101, 144, 150, 183, 198; supernatural, 33, 62–63, 81, 102, 138, 213, 156; vocabulary of, 13, 131

original acquisition of concepts, 8–9, 78, 80–81, 83–85, 107, 109, 113

ostensive proofs, 208

ovist theory of generation, 31, 33, 82, 156, 169, 171–72, 180, 187, 199

palingenesis, 201

panorganicism, 181

pantheism, 212

parthenogenesis, 32, 170–72, 212–13

perfectibility, 105, 111, 202, 210

Perrault, Claude, 33, 169

Persius, 149

Philanthropinum School, 95, 105–6, 153. *See also* Basedow, Johann

phyletic lines, 11, 14–15, 98, 100–101, 105, 200

physicotheology, 61, 102–3, 216

physics, 6, 8, 33–36, 39–40, 44–45, 58, 100, 115, 123, 172, 185, 225

physiognomist of reason, 205

physiological, 38, 41–42, 68, 101, 113–14, 119, 121–22, 124, 132, 175, 202

physiologist, 42, 58–59, 104, 122–23, 132

physiology, 35–36, 38, 57, 100, 113–15, 119–24

Piché, Claude, 158–59, 204, 211, 214, 224, 234

Pillars of Hercules, 147–48

Plato, 33, 74, 81, 83, 89, 109, 122, 126, 142, 147, 171–72, 192–93

pneumatism, 82, 109

poetry, 1, 155, 211

polydactylity, 4, 97, 180
polygenesists, 10
polyp, 3, 42–43, 156, 178
predispositions, 7, 11, 13, 104, 106–8, 136, 153, 160. *See also* disposition; germs
preexistence theory of generation, 3–5, 11, 13, 22–23, 28, 32–34, 37, 43, 62–63, 72, 81–83, 97, 99, 102, 112, 156–57, 160–61, 169, 172, 178–80, 187–88, 194, 199–202, 221
preformation theory of generation, 6–8, 32–33, 43, 60, 62, 81, 89, 99, 108, 138, 144, 156–60, 169, 171, 178, 187, 204, 213–14, 221
Preus, Anthony, 157, 234
producta, 109
propagation, 19, 25, 50, 71, 99, 105, 130, 187–88
psychologism, 205–6
psychologists, 111, 113, 121–22, 124
psychology, 110, 113–15, 119–24, 205–7

Quarfood, Marcel, 159, 234
quid facti, 122, 131–32, 135
quid juris, 131–33

race, 10, 92, 95–96, 99–101, 103–4, 108, 151, 156, 160, 162, 185–86, 197–203, 220, 223
racial characteristics, 11, 96, 99, 101–2, 198, 200
Ramsay, Thomas, 2–3, 156, 234
rationalists, 72–73, 83
Réaumur, René, 35, 44–45, 56, 172–73, 180–81
regenerate, 3, 11, 43, 160, 178
replication, 42–43, 45, 153
reproductive imagination, 114, 118–19, 122, 207
resemblance, 49, 100, 135, 162, 170, 183, 198, 217
respiration, 30, 36, 173
Richards, Robert, 161–62, 216, 234
Robert, Jason, 158, 235
Roe, Shirley, 157, 169, 172, 186, 226, 235
Roger, Jacques. 156–57, 169, 175, 235
Romanticism, 2, 5, 162, 210, 216, 219, 225
rooted, 32, 129–30, 151
rootedness, 129
Rousseau, Jean-Jacques., 52, 73, 191, 210
Ruestow, E. G., 169, 171, 235

Ruhe family of polydactyls, 97, 180
Russian doll model of generation, 23, 62, 156

Sachs, Julius von, 163, 174, 182–83, 185
Schelling, Friedrich, 173, 210, 236
Schiller, Friedrich, 155, 236
Schlegel, Friedrich, 155, 211, 236
scholastics, 33, 58, 172
Schönfeld, Martin, 184, 186, 236
Schrecker, Paul, 172, 236
Schultz, Johann, 191, 208, 221, 236
seminal fluids, 19–22, 28, 32, 37, 44, 97–98, 103, 165, 171
Sennert, Daniel, 20, 164, 171, 220
sensibility, 9, 79, 84, 94, 109, 116–18, 129–30
sensitive, 76–79, 83, 86–87, 90, 93, 109
sexdigitalism, 186
sexuality, 37–38, 175
'sGravesande, Willem, 33–34, 172
Shaftsbury, Anthony, 73
Shaw, Brandon, 158, 214, 236
Shell, Susan Meld, 200, 236
shell formation, 173, 211
skepticism, 10, 13, 21, 51, 53, 73, 79, 87–88, 109, 126, 130, 133, 147
slaves, 179, 203
Sloan, Phillip, 157, 159, 163, 174–77, 179, 182–83, 186, 198, 204, 220, 223, 226
soul: Aristotle on, 7, 17–18, 65–66; and body, 8, 30, 71; immortality of, 64–65, 121; and innate knowledge, 32, 58, 72, 74, 81, 88–89, 108; as principle of life, 8, 17, 86, 94, 99; and psychology, 82, 111–12, 115, 120, 123; as source of organization, 5, 8, 17–18, 29–31, 45, 66
spermatic animalcules, 3, 31, 33, 156, 169, 171, 179
spermatozoa, 22, 29, 169
spermist theory of generation, 82, 156, 180
spontaneity, 12, 108, 111, 116–18, 196
spontaneous generation, 7, 82, 158, 162, 165, 212, 224, 227, 229
Stahl, Georg, 53, 65, 69, 157
Stahl, Peter, 20, 165, 234
Stillingfleet, Edward, 23, 166
Sturm, Thomas, 206, 237
subreption, 70, 75–76, 81, 86–87, 95–96, 99–100, 106, 113, 159, 190, 216, 220

INDEX

subreptive axioms, 70, 75, 84, 87, 101, 120–21, 190
surreptitious concepts, 70–72, 79–80, 84, 95, 189–90, 193
Swammerdam, Jan, 29–31, 169
Sydenham, Thomas, 21–22, 166
synoptic view of reason, 139

Talbott, Stephen, 160, 237
taxonomical, 4, 18, 20, 24, 96, 100–101, 109, 137, 142, 182
taxonomists, 4, 16, 18, 20, 24, 163
taxonomy, 3, 5, 16, 18, 20, 22, 36, 47, 100–101, 113, 138, 182, 199, 206
teleological, 104, 106, 128, 140, 143–44, 150, 161, 170, 211, 215–16
teleology, 2, 14, 104, 106, 142, 153, 200, 215–16, 225
teleomechanism, 215
telic course of development, 52, 140, 144
Terrall, Mary, 179, 181, 184, 237
Tetens, Johann, 110–15, 118–19, 123–24, 132, 160, 201, 204–7, 237
theology, 33, 61, 114, 145, 162, 212, 223
Tonelli, Giorgio, 179, 184, 192, 237
topography, 140
topology, 46, 68
transformist, 198
transmigration, 31
transmutationist, 198
transplanting of species, 98, 101, 103
tree egg, 37
Tree of Diana, 44–45, 179
Trembley, Abraham, 3, 43, 156, 178, 181, 237
trunk, 42, 49–50

Uebele, Wilhelm, 204–5
unfolded, 11, 103–4
unfolding, 6, 11, 13, 62, 102–4, 108, 150, 158, 160, 178, 187, 200–201, 210
unity: of apperception, 12–14, 94, 108, 118–19, 130, 133, 137–41; of cognition, 12, 94 107; of experience, 12, 93–94, 108; of judgment, 12; of matter, 29–30, 164; of nature, 14–15, 17, 41, 50, 61, 102–3, 106–7, 142, 149, 150, 152; of organisms, 9, 11, 19, 25–26, 30, 64, 83, 128–29, 211; of phylum or species, 14, 104, 161–62; of reason, 9, 12, 92, 94, 129, 131–35, 137, 149, 193, 196; of rule, 10, 12–13, 93, 136, 197; of system, 128, 142, 145, 209

variation, 4, 16, 48, 82, 97–99, 160, 200, 225
varieties, 4, 61, 66, 96–98, 101, 103, 127–28, 130, 180, 197–98, 202, 220, 222
vegetative force, 66, 130, 153, 166, 216
virgin birth, 32, 81, 171
vital force, 36, 43, 69–70, 72, 160–61, 171, 203, 215–16
vitalism, 1, 5, 65, 69, 171
vocabulary of life, 13, 89, 127, 131, 141, 159, 195, 199, 204, 212, 216

Walmsley, Jonathan, 165–66, 220, 237
Warda, Arthur, 185, 237
Watkins, Eric, 184, 195, 204, 206, 225–26, 237
Westfall, Richard 163–64, 237
Wilkie, John 157, 237
Winter, Alois, 186, 191, 237
Wolff, Casper Friedrich, 5, 36, 60, 112, 119, 157, 161, 188, 201, 205, 216–17, 238
Wolff, Christian, 52, 54, 57–58, 64, 77, 126, 175, 182, 185, 189–90, 192, 237–38
womb, 19, 150, 152
Wubnig, John, 159, 238

Zammito, John, 159, 181, 189, 192, 197, 238
Zöller, Günter, 158, 199, 210, 215
zoology, 100, 175
Zweig, Arnulf, 206, 238

www.ingramcontent.com/pod-product-compliance
Lightning Source LLC
Chambersburg PA
CBHW021941290426
44108CB00012B/915